Harnessing Solar Power

The Photovoltaics Challenge

Harnessing Solar Power

The Photovoltaics Challenge

Ken Zweibel

Plenum Press ● New York and London

Library of Congress Cataloging-in-Publication Data

Zweibel, Kenneth.
    Harnessing solar power : the photovoltaics challenge / Ken
  Zweibel.
        p.   cm.
    Includes bibliographical references and index.
    ISBN 0-306-43564-0
    1. Photovoltaic power generation.   I. Title.
  TK2960.Z945 1990
  621.31'244--dc20   90-39905
                                                          CIP

ISBN 0-306-43564-0

© 1990 Ken Zweibel
Plenum Press is a Division of
Plenum Publishing Corporation
233 Spring Street, New York, N.Y. 10013

Printed in the United States of America

To my wife, Rebecca,
and my children, Stephen and Jane,
with love and respect

☀ Preface

Photovoltaic (PV) devices transform sunlight directly into electricity. They have been around for about thirty years, doing exotic things like generating electricity for satellites in space. But the real value of PV lies in its potential to produce electricity on Earth, and to do so cheaply enough to compete with conventional sources of electricity like nuclear, coal, oil, and natural gas. Even though PV is still much too expensive to compete directly with conventional electricity, research over the last fifteen years has moved it markedly closer to this goal. When PV is sufficiently developed, it should become a new means of providing a large fraction of our electricity, and in a way that will minimize environmental stresses. The seeds of this eventual large-scale use of PV have already been sown.

This book examines the potential of PV and the barriers that stand in its way. Since PV is a fairly new way of producing energy, a number of myths surround it. For instance, many people feel that the need for large tracts of land is a major obstacle to the future use of PV. But in reality, PV uses about the same amount of land as needed for mining and then using coal for electricity. This book devotes a great deal of attention to correcting the misconceptions that pervade people's understanding of PV.

The real barriers to the adoption of PV on a global scale are technical. These barriers directly relate to the cost of PV-generated electricity. The active layers in PV devices—the layers that transform sunlight into electricity—are based on so-called semiconductor materials, which are similar to the materials used in computer chips. The cost issues of PV are associated with *manufacturing these high-quality semiconductors on an unheard-of scale,* dwarfing the scale of the existing computer industry. But PV's tradition of steady technical progress gives us a good deal of confidence in its future. Much of this book is about the great track record of the community of PV scientists who have consistently outperformed skeptics.

The technical barriers will eventually be overcome. The large-scale impacts of PV are just a matter of time ... but also a matter of who will dominate the expected multibillion-dollar PV industry. Will it be the US, the Germans, or the Japanese? Will we be importing our PV, as today we import oil? For example, 1989 saw a deal in which the key US company in PV—ARCO Solar—was sold by Atlantic Richfield to a German electronics giant, Siemens. ARCO Solar was not only the world's largest PV company in terms of sales, it also dominated research in several of the key PV technologies that are just now beginning to have an impact. The issue of the US role in PV may eventually come to be the focus of discussions about PV, as other issues such as land and technical barriers recede. Before we lose out completely in PV, we as a nation need an energy strategy that reflects the importance of PV and the importance of *an American role* in dominating its development. Business-as-usual will not get the job done. Examples abound in this book of the failure of existing US corporate and federal policies concerning this issue.

In the final chapter of this book, we outline a strategy for assuring the progress of PV in America. The nation will need two things: An enlargement of the federal PV research budget from its current shoestring level (under $35 million annually) to a reasonable level ($100 million annually), and the stabilization of the US PV industry. At the moment, the US industry leads the world in PV technology, but many PV companies are in imminent danger of either being sold to foreign buyers (at 10 cents on the dollar) or going belly-up. The cause is the shortsightedness of our corporate and federal decision-makers, who have almost without exception emphasized near-term goals over longer term values. Stabilizing our PV industry is a complex matter, but perhaps could be accomplished through environmental regulations mandating the increased use of PV. For instance, utilities or municipalities that failed to meet stringent environmental limits for various emmissions (e.g., sulfur dioxide, carbon dioxide, particulates) could offset penalties by installing PV. For the first decade of such a program, mandated projects could be small and be subsidized with low-interest loans from the government. PV electricity costs are very sensitive to interest rates. Dropping rates in half drops PV costs in half. To assure ongoing technical progress, subsidies for each new PV project should be incrementally reduced from previous projects during the course of the program. This would force PV manufacturers to stay on track in terms of continually reducing costs. This kind of continued price discipline is essential to success.

Such a program could be executed on a very small scale (less than $5 billion in the critical 1990s) and still be enough to secure America's position as a key participant in PV during the 21st century. But no one would want to engage in such a program unless they felt convinced that PV could someday be cost-competitive. Similarly, they would need to be sure that low-cost PV would eventually have a major impact on energy and would help to improve the environment. It is our hope that this book will assist in answering positively the question of PV's value and prepare for a more substantial national commitment to PV during the 1990s. *The PV challenge* is to rapidly develop this exciting new energy technology and to do in a way that allows the US to play a significant role in its future.

It would be remiss of me not to state that the opinions in this book are those of the author. They in no way are meant to represent the views of the author's employer, the Solar Energy Research Institute (SERI). The Midwest Research Institute manages SERI, and that organization is also not responsible for the views expressed here. Finally, SERI is funded in large part by the Department of Energy. Again, its opinions and those expressed here may not coincide.

# Contents

1 ☼ Electricity without Pollution

Introduction

Many people feel that when other forms of energy are exhausted or become too environmentally costly, solar energy will replace them. But are they being realistic? Very little information is commonly available to support their idea. In fact, the concept of a large role for solar energy is usually stated as an antidote to an all-nuclear future by those who seem allergic to that possibility. As such, being optimistic about solar energy may seem naive or even deceptive. In fact, most energy experts do not expect that solar energy will provide a substantial fraction of our energy in the 21st century. Yet that is the thesis of this book.

Photovoltaics (PV)—the direct conversion of sunlight into electricity—is a solar technology with the potential to supply essentially all the energy we need. Its first uses have been tiny and remote—as remote as outer space or the Himalayas—but now PV is coming down to Earth, and PV costs are plummetting, too. Unfortunately, this nascent revolution in energy production is happening in relative anonymity. The public has hardly heard the word *photovoltaics;* government and corporate decision-makers underestimate its progress and potential; and most US investors ignore it. But if its technical progress continues, PV could be recognized as the preferred method of providing new electricity in the next century.

PV depends on solar cells, which are layered devices designed to turn sunlight into electricity. Leave them outside, and they will make electricity as long as the sun shines. Solar cells have been around since the 1950s, but the first rush of enthusiasm for developing them for energy conversion on a large scale occurred about fifteen years ago, with the oil crisis of the mid-1970s. Public interest in PV soared. Significant corporate and government programs began.

Since then, nearly a generation of scientists have contributed to the growth of PV. More than a billion dollars in research money has been spent. Yet where are the PV panels that would provide billions of kilowatts of power? That is the uppermost question, as far as the public is concerned. If they think of PV at all, they must think it a failure — a forgotten, deemphasized remnant of the last energy crisis.

This is not the place for a dissertation on the realities of high-risk, long-term research — or the commitment that fosters its success. The truth is that while the public has been otherwise occupied, PV *has* matured and could soon become a major contributor of world electricity.

Where We've Come from

The principles behind PV are by no means a recent discovery. The French scientist Edmund Becquerel discovered the PV effect in 1839. He experimented by placing two electrodes in a beaker full of fluid. A spontaneous electric current flowed when the beaker was exposed to sunlight. This current must have seemed very mysterious to him. He and his contemporaries simply did not have the theoretical understanding of physics and chemistry to comprehend the PV effect. In fact, as scientists now know, PV is a complex, quantum mechanical phenomenon. Becquerel and other 19th century scientists had no idea that they had discovered a means of transforming light energy directly into electricity. PV languished while the theoretical framework for explaining it was developed. For much of the 19th century, the PV effect was almost ignored.

One exception to this was the work of Willoughby Smith, who discovered in 1873 that the element selenium was light-sensitive. He found that selenium conducted far more electricity when it was illuminated than it did when it was in the dark. This phenomenon became the basis of selenium light meters for photography, used as early as the 19th century. Later, the English scientist William G. Adams proved that light could generate an electric current in selenium.

Perhaps the quintessential idea behind interest in PV is that of using it on a large scale to generate electricity. Charles Fritts, an American who invented the first practical solar cell—made from selenium—speculated that PV "may ere long compete with the dynamo-electric machine." He wrote in 1886 that "the supply of solar energy is both without limit and without cost" and thus should be available long after we have run out of fossil fuels. It is amazing that this ambitious idea of how to use PV goes back to this early origin.

But scientists of the time were used to the thermal generation of electricity achieved by the burning of fossil fuels. They either ignored Fritts' work or considered it a fraud—because, as they thought, it seemed to violate the principle of conservation of energy. They did not understand that the energy produced by PV was coming from the energy of sunlight.

It was only during the 20th century, with the development of a far deeper understanding of the nature of matter (quantum mechanics and solid-state theory), that the PV effect became comprehensible. With that improved understanding, the idea of harnessing PV was resurrected. By 1930, the selenium cell had been rediscovered, and thoughts of limitless power were again being considered. In 1931, *Popular Science* speculated that if PV cells were fully developed, they might power "a huge solar electric station at a cost no greater than would be required to build a hydro-electric station of the same capacity."

But selenium cells were not very effective in changing light into electricity. The best they could do was to transform about 1% of the incoming sunlight into usable electricity. This fractional amount, called the cell's efficiency, was much too low to be economical, given the cost of making the cell.

The Birth of Practical PV

Many consider work during the 1950s at Bell Laboratories to be the real birth of PV. PV cells are actually quite similar to several of the key devices—rectifiers, diodes, transistors—behind the solid-state electronics revolution of the second half of the 20th century. This is the revolution that brought us the transistor radio, the tubeless television, the stereo amplifier, the mainframe and home computers, lasers, and modern telephone systems. The key electronic devices for this revolution were being developed in the late 1940s. Accelerated progress in PV took place in the 1950s because the science needed for PV was very similar to the science of solid-state electronics.

A Bell Labs scientist, Cal Fuller, was working on a project to perfect rectifiers made of silicon. Silicon, which is one of the Earth's most abundant elements (it is the main constituent of dirt), is the semiconductor material that has since become the mainstay of the electronics industry. The silicon rectifiers that Fuller was working on were devices that allowed electric current to flow in only one direction. In fact, they are close cousins of PV cells, which are themselves really just light-activated rectifiers. A co-worker of Fuller's, Darryl Chapin, improved the effectiveness of the silicon rectifier by adding impurities that strengthened its electronic properties. Then, Fuller's supervisor, Gordon Pearson, exposed one of the rectifiers to light, and it generated electricity.

Scientists on another Bell Labs project were trying to develop a power source for remote communications systems. They had tried using selenium solar cells but dropped them because of their low efficiencies. Cal Fuller, Darryl Chapin, and Gordon Pearson joined the program to produce remote power and soon had made a silicon cell that was able to convert 4% of sunlight into electricity, a fourfold improvement over selenium cells. By May 1954, they had reached a 6% conversion efficiency.

Their achievement caused quite a stir—illustrative of the excitement that can be generated by the idea of a clean, cheap, inexhaustible source of energy. We have had a recent (1989) example of the same thing in the meteoric excitement caused by the University of Utah's announcement of "cold fusion." Pictures of Chapin, Fuller, and Pearson with a solar-powered transistor radio created headlines all over the US and even internationally. Chapin recalled that their discovery "shook the whole world." *Business Week* predicted that PV would soon power phonographs, fans, TVs, and lawn mowers. But Fuller and

Chapin took a more modest view. "We tried to avoid making too much claim for it," Chapin has said, "because we knew it was in the laboratory stage, it was an expensive process, and there was a lot to be done before we could speak of lots of power."

The Bell Labs work also marked the birth of the break between overblown public expectations for PV and the very real, but unspectacular technical progress needed to develop PV for practical uses. The public enthusiasm of the 1950s was way ahead of the technology. This kind of phenomenon has occurred several times since in the history of PV, most notably in the late 1970s. Sometimes high public expectations have actually helped PV—when popular enthusiasm has triggered greater efforts to be made. But unrealistic expectations have also hurt PV, when the public's patience has run out and been replaced by a backlash of disappointment. PV technology is much more complex and requires a far more sophisticated technical foundation than the public realizes.

Fuller and Chapin eventually reached 10% sunlight-to-electricity conversion efficiency. They also established their material—silicon—as a key PV material, one that is still dominant today. But hope of using PV on a large scale soon faded as practical barriers stalled progress. Compared to conventional electricity in the 1950s, the cost of PV-produced power was a thousand times too much.

But another, seemingly unrelated matter—the US space program—helped PV get going again. PV cells weigh very little in terms of the power they produce. They were a natural for use in space, where payload weight is critical. It is simply impossible to send most fuel-driven energy sources into space—and then send up more fuel to power them. One of the first satellites, Vanguard I (March 1958), used silicon PV cells to power a radio beacon. The satellite designers did not have an ambitious project in mind; all they wanted was a radio signal indicating that the satellite was successfully in orbit. However, they forgot to provide a turn off switch, and for eight years Vanguard's PV-driven radio beacon filled the airwaves.

The reliability and lightness of PV so impressed NASA that they adopted PV to power almost all of their subsequent satellites. Nearly every communications satellite, military satellite, and scientific probe is powered by PV. In space, PV *is* the conventional power source and is likely to remain so for the foreseeable future.

The PV Industry

A small PV industry capable of manufacturing solar cell products was born out of the need to supply the space program. During the 1960s, companies such as Hoffman (now Applied Solar Energy Corporation), Heliotek, International Rectifier, RCA, and Texas Instruments produced efficient, dependable, though expensive cells for space. In 1970, sales were about 80,000 watts (80 kilowatts) annually at an average cost of about $150 per watt. This cost would translate to electricity at about $30 per kilowatt-hour if such PV cells were used on Earth, about 600 times more than we spend on conventional electricity. But for space, cost meant little; durability, efficiency, and lightness were the overwhelming priorities.

The "far-out" idea of using PV for cheap, terrestrial power languished during the 1960s. But a sophisticated understanding grew of PV materials and devices. Meanwhile, the use of PV in space demonstrated that cells were extremely rugged, capable of great dependability despite long exposure to vacuum, radiation, and temperature extremes. Satellite manufacturers using PV could tolerate high cost because the cost of PV was a small fraction of their total cost. But they could *not* tolerate an undependable power source which would endanger their entire investment. For three decades, PV has proved its dependability in space. As far as NASA is concerned, dependability has become a byword of PV.

The Energy Crisis

The next significant influence on the evolution of PV was the energy crisis of the 1970s. The price of gasoline escalated from under 30 cents a gallon to over a dollar a gallon. People felt that energy was in tight supply and would soon run out. They feared that our dependence on OPEC for oil threatened our national security. No one who experienced it will ever forget the lines of motorists waiting for gasoline, overwhelmed by the helpless feeling of being controlled by unfriendly foreign interests. Despite PV's astronomically high cost, it was seriously considered as a possible energy option. A major conference was convened by the National Science Foundation at Cherry Hill, New Jersey, in October 1973, to examine the issues and potential of PV. A research agenda was developed. Government programs were started through various agencies, culminating in work at the Department of Energy (DOE) that is still continuing today. PV funding at the Department of Energy eventually reached an apex of over $150 million annually in 1980. It has since dropped to about $35 million. Still, much of the progress in PV occurred because of the research agenda started at the Cherry Hill conference in 1973.

Hopes and Disappointments

Skeptics might be tempted to consider the history of PV a chronicle of repeated hopes and disappointments. Certainly PV has not yet become a significant provider of electricity to anyone. They might think that those who suggested PV as part of a solution to the energy crisis in 1973 were being too credulous.

But the idea of using PV and sunlight for producing electricity is a good one. The simple fact is that in the past, technical barriers to PV were enormous and repeatedly underestimated. During the 19th century, the obstacles were insurmountable. Scientists did not even know what a semiconductor was or how semiconductors could be made and exploited for electronic devices. They were simply unprepared to solve the problems associated with developing PV for practical use. But subsequent scientific progress—in theoretical physics, electronics, and especially in semiconductor materials processing—has provided a base for real progress in PV. Scientists can now draw on this technical base to improve PV devices and then manufacture them cheaply. Rather than being a history of failure, PV's development is a true reflection of the complexities of the evolution of a significant new technology. Other technologies that are just now emerging—materials that are superconductors at higher temperatures, fusion, genetic engineering—will almost certainly follow similarly arduous paths to success.

Since emphasis on terrestrial PV began in 1973, the scientific community has reduced PV costs more than 20-fold. PV-generated electricity now costs about 30–40 cents per kilowatt-hour, a far cry from the costs of the 1960s and 1970s. Meanwhile, the trend toward less expensive PV continues unabated. Based on existing technical progress, the cost of PV by 2000 AD could be down to about 6–10 cents per kilowatt-hour. At that price, PV will be fully competitive with fossil fuels, *and the price of PV will still be falling* because such progress will not have exhausted PV's full potential. The cycle of hope and disappointment partially explains why PV is not given greater public attention these days. It is certainly not being ignored for a lack of technical progress. As we shall see in subsequent chapters, significant records for PV performance are being set almost daily. But the public perception of PV in the 1990s may be changing for the better.

Concern over the greenhouse effect has brought PV—which produces no greenhouse gases—back to the edge of the limelight. PV is mentioned frequently in meetings of the energy and environmental communities. Influential people are suggesting that PV can make a major contribution to solving our energy and environmental problems. Politicians are beginning to take PV seriously as an alternative source of energy. PV has come a long way since 1973, when it was just another good idea. The new ideas of the late 1970s and early 1980s are being turned into practical PV devices. A *fair* assessment of existing PV technologies would show that PV's practical potential is immense.

A Scenario for PV's Role in the 21st Century

In the 21st century, PV could provide a major proportion of our electricity while causing very minor environmental effects.

An ever-growing population and the ongoing desire for affluence are driving the world's need for energy. Conventional energy sources—oil, coal, natural gas, nuclear—pollute in ways we once could afford, when the world's population was relatively sparse and when energy-intensive economies were fully developed in the West alone. The side effects from conventional energy—the greenhouse effect, air and water pollution, supply crises, international supply tensions, resource depletion, fear of nuclear accident—will not be affordable in the more populous world of the 21st century. But PV could become a key part of the solution to our problem.

Based on the rate of present technological progress, PV should grow perceptibly (30–50% annually) in the 1990s. At the end of the 1990s, progress should become even more dramatic, providing a base for a vast expansion of PV in the first decades of the 21st century. As an antidote for environmental and population pressures, PV could then provide a measure of room for the world's economies to grow and for us to avoid draconic changes in our deep-seated expectations of affluence and consumption. PV is an example of how we can be smarter about our technologies—minimizing the waste they produce—and still sustain ambitious growth. As a bonus, PV costs should be falling well into the next century as technical progress and innovation are translated into lower costs.

2 ☼ What is Photovoltaics?

Photovoltaics (PV) encompasses the entire technology of the use of solar cells. Solar cells—or PV cells—absorb sunlight and change it directly and continuously into electricity. They do this without pollution, noiselessly and without any moving parts (although some PV *systems* have sun trackers with moving parts).

Before trying to understand the workings of a solar cell, let's examine some properties of light and the solar spectrum. Knowledge of the spectrum is valuable because the idiosyncracies of sunlight seriously influence the way PV cells are designed.

Sunlight

Sunlight is the fuel of solar cells. It carries the energy that solar cells convert to electricity. Above the Earth's atmosphere, sunlight carries over 1300 watts (W) of power per square meter, but not all of this reaches the Earth's surface (Figure 1). Reflection and absorption of light during its passage through the atmosphere are substantial. At noon on a cloudless day, sunlight provides about 1000 W (1 kilowatt or kW) of power on each square meter of the Earth's surface, 300 watts less than it had in space. It loses the 300 W during its passage through the atmosphere. A thousand watts of electricity can power ten typical (100-W) lightbulbs.

There is a difference between kilowatts, the term we just used, and kilowatt-hours (kWh). Kilowatts measure power. A kilowatt-*hour* is a measure of energy. Energy is power times the duration during

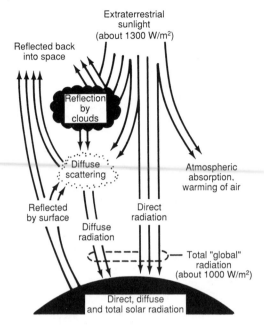

Figure 1. Not all the sunlight incident on the top of the Earth's atmosphere reaches the Earth's surface. About 300 W/m² is lost to reflection and atmospheric absorption. At noon on a sunny day, about 1000 W/m² of sunlight is available at the Earth's surface for conversion to electricity by PV.

which the power is being produced. In other words, a PV system producing 500 kW of power for 3 hours produces 1500 kWh of energy. We generally speak of kilowatt-hours when we talk about electricity. Your electricity bill is given in kilowatt-hours. A typical US family home uses about 5000–10,000 kWh of electricity annually.

Sunlight is most intense at noon. If the amount of sunlight in the hours around noon were much less, PV would be an impractical concept. There wouldn't be enough energy in sunlight to be of use. Fortunately, there is a plateau of relatively intense sunlight in the 5–7 hours around solar noon. Solar noon is the time when the sun is highest in the sky, and it is usually within an hour of 12 noon on the clock.

The total energy available from sunlight on a typical day in the US is approximately 5 to 9 kWh for each square meter (kWh/m²) of exposed area. For sunny Las Vegas, it's 9.7 kWh/m²; for cloudy Seattle, 4.78. Boston is 5.3; Washington, DC, 6; and Denver, 8.5. These numbers represent the practical extremes of solar availability in the US. Almost 90% of the country gets between 6 and 8 kWh/m² daily, which is plenty for the effective use of PV.

The amount of sunlight available in a particular place on an annual basis gives us a measure of how much energy we can convert to electricity. There is somewhere between 2 and 3 megawatt-hours (million watt-hours, MWh) per year of sunlight incident on each square meter of the US, depending on climate and latitude. In total, the US receives about 2.4×10^{16} kWh of sunlight per year, more than 10,000 times our annual use of electricity. That is clearly a huge solar resource—and a domestic one, at that. (See also Chapter 13, Sunlight, for more details.)

Direct and Diffuse Sunlight

Sunlight reaching the Earth's surface has two components—direct-beam and diffuse. Diffuse light is the portion of sunlight that has been refracted or scattered in the atmosphere before it reaches the Earth's surface. We see much of it as blue sky. Under perfect, cloudless conditions, about 10% of the sunlight is diffuse. Direct-beam sunlight is that which we see as coming directly from the sun. Although it can be 90% of the sunlight at noon under cloudless conditions, it is a smaller fraction at various times of day (morning and dusk) or when it is cloudy or hazy. Under cloudy conditions, diffuse light can be 50 to 100% of the sunlight. Even in the desert, about 20% of sunlight is diffuse on an annual basis; and in northern Europe, almost 60% is diffuse. The total sunlight, called global sunlight, is the sum of diffuse and direct components. The breakdown between diffuse and direct sunlight is important, because a few types of PV systems cannot use diffuse sunlight at all.

The Spectrum

Direct sunlight appears to be yellow but is of course actually made up of a spectrum of colors. This is unfortunate for solar cells (but not for people), because solar cells would be a lot more effective if sunlight were just one color. Under monochromatic (one-color) light, a typical PV cell might be able to convert over 60% of the light to electricity; under the multicolored solar spectrum, the same cell would be able to convert only 10% of the light's energy to electricity.

A prism can be used to separate sunlight into a spectrum of colors. The differently colored lights have different levels of energy. Blue light has more energy than yellow light, which in turn has more energy than red light. Light with energies too high for us to see is called ultraviolet light, which is known for causing sunburn. Almost all ultraviolet light is absorbed by ozone as the light passes through the atmosphere. Light with too little energy for us to see is called infrared. This is the kind of light that a greenhouse traps within it and which causes it to warm up. Much infrared light is absorbed by water vapor and carbon dioxide in the atmosphere. Some solar cells can convert to electricity light from the whole spectrum (infrared through ultraviolet); but others can only respond to a certain narrow portion of the spectrum. The distribution of sunlight in the solar spectrum is a very important parameter controlling a PV cell's performance.

Individual particles of light are called photons. Photons can be characterized in two ways: by their energy (usually given in electron volts or eV, which is the energy an electron gains when accelerated by a force of 1 volt applied for 1 centimeter of distance) or by something called their wavelength. Each photon has a well-defined energy and wavelength. Wavelengths are usually given in units of microns (10^{-6} m or 10^{-4} cm). Visible light is about 0.6 micron in wavelength, which is equivalent to saying the same photon has about 2 eV of energy. We will use both energy and wavelength when describing light.

The fact that sunlight has many colors is equivalent to saying that sunlight is made up of photons of many different energies or wavelengths. The distribution of photons in the spectrum (rather than their all being of one energy) is one of the greatest limiting factors on PV cell performance.

Now we're ready to take a look at what happens to sunlight when it strikes a PV cell.

PV Cells

Figure 2 shows the basic structure of a PV cell. (PV structures vary widely. This one is simply a handy generic model.) The top layers that sunlight encounters are purposely transparent. The outermost layer, here labeled cover glass, is an encapsulating layer to protect the rest of the structure from the environment. It keeps out water, water vapor, and gaseous pollutants, which would corrode a cell if allowed to enter during its long outdoor use.

Solar cells are intended to last about 30 years; and severe tests have been developed to see that they achieve this extended life. The cover glass is often hardened (tempered) to protect the cell from hail or wind damage. One challenge that encapsulants must pass is the impact of baseball-sized hailstones moving at 52 mph. Another test involves submerging the whole structure in water without causing any damage.

After light passes through the cover glass and the transparent adhesive that holds it to the cell, it encounters something called an antireflective (AR) coating. This coating is a transparent layer designed to reduce the amount of reflected sunlight. Bare cells can look mirror bright and might reflect about 30% of the sunlight. Since the power output of a cell is proportional to the amount of sunlight that it absorbs, such losses would be intolerable. But reflection losses can be reduced by the judicious choice of an AR coating. Reflection of light passing through an interface is greatest when there is a large mismatch between the index of refraction of the two neighboring materials. A material's index of refraction is a measure of its tendency to bend light as the light enters it. Air has an index of refraction of 1; silicon's index is about 3.5, which is a large mismatch. To minimize reflection off a silicon cell, one can choose an AR coating such as zinc sulfide with an index of refraction about halfway between silicon and air. Such an AR coating can reduce sunlight reflection losses to below 5%.

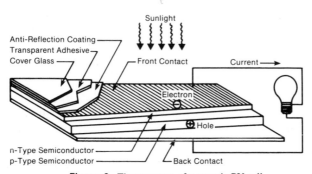

Figure 2. The structure of a generic PV cell.

Electrical Contacts

Let's return to Figure 2 to follow the journey of sunlight into a solar cell. The next layer (numerous parallel lines in the figure) is called the front contact. It is a contact—like the plug in a wall socket—between the solar cell and an external electric circuit. Its function is to carry away the electric current produced by the cell. Often it is made of a metal such as aluminum or silver.

Light-generated current flows out of a cell from all parts of its top surface. So it's important that the top contact reach everywhere on the cell. Resistance losses would mount up rapidly if electric current had to travel very far sideways across the top of a cell to reach the top contact. Logically, one would like to cover the whole top of a cell with metal to minimize this loss. Unfortunately, then the cell wouldn't work at all. All the sunlight would be blocked by the metal. So top contacts are usually made as thin fingers or grids that reach most of the cell but only block a small portion of the light. Figure 3 shows an example grid design. The long, thin metal fingers reach every area of the cell's top surface but block only a small fraction of the light (about 10%).

Actually, this is not the whole story by any means. In fact, several very interesting materials that are highly transparent, such as metal oxides (e.g., zinc oxide and tin oxide), have been found that can mimic the role of metals as far as conducting electricity. These highly conductive *and* highly transparent materials can be coated over the entire top of a cell without blocking sunlight. By this means, the top grid can be eliminated. Metal oxides are used extensively in some advanced PV designs.

Returning to Figure 2, the cell's bottom layer is called the back contact. Unlike the top contact, it can be a sheet of metal since it isn't in the way of any sunlight. Electric current flows vertically through a cell to the bottom contact. The back and front contacts are necessary bridges to an external circuit.

The middle layers between the two contacts are the critical layers. They are the guts of the cell. You will note, however, that on the way into our generic cell we have already lost about 15% of our available sunlight to surface reflection and shadowing by the grid. Fortunately, some advanced designs (such as those with transparent front contacts) do a lot better, losing as little as 5% to these parasitic processes.

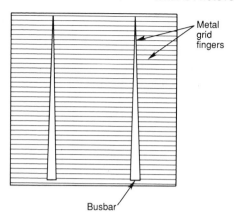

Figure 3. Schematic of a metal top grid that shadows about 10% of a cell's surface. Note that the two vertical grid fingers taper as they get further from the busbar, which is along the bottom. There is less cumulative current at the top of the grid, so the metal fingers can be thinner without incurring extra resistance losses.

The "Guts"

The two middle layers, shown as n-type and p-type in Figure 2, are where sunlight is absorbed and electricity is generated. The absorption of light and the production of electricity by these layers are quite complex. In outline, however, the process is pretty simple. Basically, light comes in; and electricity comes out. But how?

The two middle layers are semiconductor materials. Semiconductors are special electronic materials used in computers and other electronic devices. Most of the electronic consumer items (e.g., stereos, TVs, computers, VCRs) that we are used to depend on semiconductors. *Semiconductors* are named for the fact that they conduct electricity poorly in comparison to metals but very well when compared to insulators such as rubber or wood: they are *semi* (or partial) conductors. In PV cells, they have two properties that are at the heart of a cell's ability to make electricity:

1. Electrons (the constituents of electricity) are freed in a semiconductor when photons of sufficient energy are absorbed within them; and
2. When dissimilar semiconductors are joined at a common

boundary, a fixed electric field is usually induced across that boundary.

Scientists have spent decades working with these properties, trying to make the best possible PV devices. The simple version of how solar cells work goes as follows (Figure 4). Light enters the cell and is absorbed in the semiconductor sandwich (the n- and p-layers in Figure 2). Each photon absorbed in the semiconductor frees an electron (principle 1). Principle 2 says that there is an electric field at the boundary between the two different semiconductors. Electric fields push electrons. If the electron freed by light is close enough to the boundary, the electric field will push it over to the other side (say to the top) of the cell. This is called charge separation and is the essence of the PV effect.

The movement of an electron into the top semiconductor layer creates a charge imbalance in the cell (one electron too many in the top of the cell; one too few in the bottom). The extra electron in the top layer wants to reestablish charge neutrality, but it won't be able to go back across the boundary to the bottom because of the opposing electric field that sent it across in the first place. But it can take the long way around through an external circuit, where it can perform some useful work for us, and that is what it does.

This happens continuously as sunlight is absorbed in the cell. Light comes in and gets absorbed. Electrons get freed and pushed to the other side by the electric field. They pass through an external circuit and do some work for us before returning to the side of the cell from which they came. From an observer's standpoint, nothing seems to happen except that the cell makes electricity as long as sunlight shines on it.

How come the cell never wears out or gets used up? The sunlight provides the energy input. The electric field within the semiconductor sandwich is unchanged; it just exists. It acts like a hill, ready to propel electrons "downhill" should they get close enough to the slope. In principle, a solar cell could last just about forever, and were it not for relatively sensitive parts like the metal contacts or plastic encapsulants, it would.

There are some items in Figure 2 still to be explained. The semiconductors are labeled "n-type" and "p-type" because of the way they are different, which determines the electric field between them. The

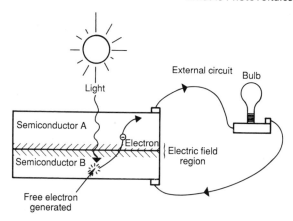

Figure 4. The photovoltaic effect (simple version). Sunlight is absorbed, generating free charges. The electric field forces one type of the charges (in this case, the electrons) to the other side of the cell, where they travel through an external circuit.

"n" stands for "negative"; the "p," for "positive." Besides the "electron" in Figure 2, there is a "hole," which just happens to be a mobile positive charge (a sort of antielectron). It works just like an electron but does everything in the opposite direction.

Putting Cells Together

A single cell by itself is almost a technical curiosity. It cannot make sufficient power to be of interest (except for powering microelectronic products such as wristwatches and calculators). A lot of cells have to be put together into modules and then assembled into arrays of modules to make electricity on a grand scale. Modules (many interconnected cells) can range from 1 square foot to 20 square feet (or more) in size. Arrays are just interconnected modules supported by some structure that points them toward the sun. If modules aren't pointed directly at the sun, they intercept less energy and reflect more of it away.

For optimal performance, the arrays can track the sun. There are two kinds of tracking systems: those that always follow the sun's path across the sky (called two-axis tracking because the tracker can move freely in two dimensions) and those that follow it in only one dimension (one-axis tracking). These latter get almost as much sunlight as the two-axis kind but cost less to build and maintain. Given existing technology, they tend to be favored unless there is a special need for tracking.

The least expensive kind of support structure for PV modules is the fixed array. It is a tilted surface that just sits there, pointing at some compromise angle (to the south in the Northern Hemisphere) that maximizes average exposure. When cost is the overriding concern, fixed arrays tend to be preferred.

Trackers intercept a fairly large amount of sunlight (20–40%) that fixed arrays cannot. However, both complexity and cost increase with greater tracking capability. As with many other trade-offs with PV, it is cost versus energy output that is at stake, and the answer usually depends on the specific technology employed.

Efficiency

A key measure of a cell's ability to produce power is called its efficiency. We discussed the meaning of efficiency previously when we described the early progress from selenium cells to silicon cells: selenium cells were 1% efficient, silicon as high as 10%. Efficiency is power-out divided by power-in. For a solar cell, this is the ratio of the electric power produced by the cell at any instant versus the power of the sunlight striking the cell. The same measure, also called efficiency, characterizes the energy output of cells, modules, or arrays versus the amount of available sunlight during some period of time — an hour, a day, a year, or the lifetime of the device. Different solar cells (or modules, or arrays) have different characteristic efficiencies. Higher efficiency is obviously better if costs are the same. Efficiencies will not fluctuate much over the life of a cell unless the cell is degrading.

By definition, the higher a PV device's efficiency, the more electricity it produces for a given exposed area. Light-to-electricity conversion efficiencies of commercial PV *modules* (large area devices) are between 5 and 14%. (Experimental cells can be higher—some over 30%.) At 5 to 14% efficiencies, PV modules produce between 50 and 140 W from 1000 W of sunlight (output power equals efficiency times available sunlight). Running for a typical day outdoors (let's say without tracking, in Washington, DC, where there are 5 kWh of available sunlight per m²), they would produce daily about 0.25 to 0.7 kWh/m² (efficiency times total energy).

PV and US Electric Demands

To get some perspective on this output, let's consider the energy use of a typical family. Assuming that an average family home uses 600 kWh of electricity each month (20 kWh/day), let's try to meet this load (on average) with 10%-efficient PV modules. One square meter of modules located in Washington, DC, produces an average of about 0.5 kWh each day, or 15 kWh each month. We want 600 kWh, so we need 40 m². In more familiar units, that is about 400 square feet, or about 20 feet on a side. The size of the roof over the master bedroom? Thus, a fairly typical US electric load for a single-family home could be offset by a reasonably sized roof area devoted to PV modules. Actual roof area requirements will be more because modules cannot be perfectly packed together and the angle of the roof may not be optimal in terms of intercepting the sun's rays.

Let us consider a much more ambitious scenario. The US now uses about 2.5×10^{12} (2500 trillion) kWh of electricity annually. This is *all* the electricity that our utilities provide. We do not propose that *one* PV system would be a good way to meet this load, but it's possible. And it is quite illustrative to consider what it might be like.

PV on a Giant Scale

We want to know how much land would be required for PV to produce the same amount of electricity as the US consumes on an annual basis. We want PV electricity to match the actual total US electricity production. Figure 5 is a map of the US showing approximately how much sunlight is available on an annual basis *for a fixed flat-plate array*. The information on this map is extremely important, because it not only defines the best PV sites but also gives the possible PV energy output of every region of the country. (We look at the distribution of sunlight—as represented by this map—in more detail in Chapter 13.) Note that we have chosen a fixed array and not a tracking array, which would get substantially more sunlight. But two-axis arrays are not necessarily more cost-effective than fixed arrays, and they require about three times more land to avoid self-shadowing of one tracker by another.

Let's choose Missouri for our all-encompassing fixed flat-plate PV installation. Missouri is an average site in the US, in the sense that it gets about an average amount of sunlight—some 1800 kWh of sunlight per m^2 annually. Let's assume a 12% sunlight-to-electricity conversion efficiency for our PV system. This is an ambitious assumption, but one within the range expected of PV. Another factor increases land area: the packing factor of the arrays. The packing factor is the ratio of array area to the actual land area. If arrays are tracking, they need substantial space between them or else they'll shadow each other. Nontrackers use less space. Let's assume a packing factor of 50% (i.e., as much unused land as array area).

Using these assumptions, a square meter of land would produce

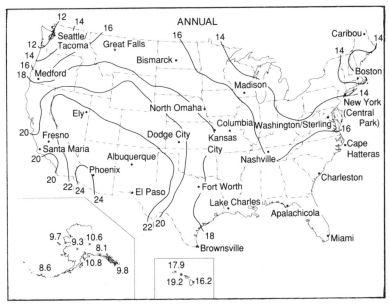

Figure 5. The United States is blessed with enough sunlight to make PV practical throughout the country. About 90% of the nation receives within 25% of the amount received in an average location like Kansas City, Missouri. The map shows annual total sunlight in units of 100s of kWh/m^2.

about 108 kWh (12% × 1800 kWh/m^2 × 0.5) of electricity each year. That means that 23,000 km^2 (8900 mi^2) would produce the total amount of US electricity. This is a circle with a radius of 86 km, or 53 miles. This may sound like a lot of land. But in certain places in the US—and often ones that are sunnier than Missouri—it's not. For instance, the Mojave desert is 34,000 km^2 in area, one and a half times the size needed to produce all of our electricity. Yellowstone Park is over 8000 km^2; and the Air Force owns a bombing range in Nevada that is over 16,000 km^2 in area. We already have socially acceptable uses for such large land areas; providing all of our electricity is clearly another important and socially acceptable goal, especially if we chose land (unlike in Missouri) with few competing uses.

Let's be very clear: We are not suggesting that we turn Yellowstone or any of these particular areas into PV sites. We are saying that land of equivalent size can certainly be found for PV.

Here is another example that is more directly related to energy production. People aren't used to thinking of setting aside 23,000 km^2 for energy production, but in fact we're already using similar areas for that purpose. Quebec Hydro, which exports hydroelectric power to the northeastern US, has about 25,000 km^2 of reservoirs behind their dams, and they have only about 50 billion W of potential capacity. Run 24 hours a day, year-round (which is impossible, but illustrates an extreme case), such a hydroelectric capacity would provide only a fifth of the US electricity requirement. But properly placed in the US, that same area devoted to PV would produce *all* of our electricity.

Offsetting a Nuclear Power Plant

Carrying the example of one huge installation to extremes is probably not worthwhile. We're much more likely to construct many smaller PV systems in order to minimize transmission losses from the source of the electricity to where it is needed. So let's see how much land is needed for a smaller PV system, one able to provide the same electricity as a nuclear power plant producing about 6 billion kWh. This is very much like a typical nuclear plant, which is capable of 1 billion W of peak capacity operating at 70% annual capacity. It is also a large, but practical scale for PV.

With the same assumptions (same Missouri location, etc.), we find that 60 km^2 (23 mi^2) of PV would provide the energy of a nuclear plant. This is a circle 4.4 km (2.7 mi) in radius. That's only 4400 m in radius! To offset a nuclear power plant—with one kilowatt-hour of PV for each kilowatt-hour produced by fission. This kind of land might be available at, say, the new Denver airport planned for the 1995 time frame. Denver has purchased about 180 km^2 of land for the airport. Useless scrubland between runways would certainly be enough for a few billion kilowatt-hours of PV electricity.

Comparing PV, which generates electricity only during the day, with a nuclear power plant (which can generate electricity at any time) is not so clear-cut. To be fully comparable, an electric storage method has to be assumed for PV. This adds cost to PV electricity and also increases land requirements if the storage is not perfectly efficient. Later on, we will look at PV and storage in much more detail and see that the comparison of PV and nuclear power can still be a favorable one for PV.

If a skeptic were to think that the land area assumptions made here are too optimistic, he could double the radius (4 times the land). For the nuclear power plant, that would be a radius of 8800 m. That still isn't a lot of land area to produce the same output as a nuclear power plant. This simple analysis shows that total land area is not a critical problem with PV in the US. In fact, a recent analysis by the Department of Energy states that the land for conventional electricity generation by coal combustion, when land for mining is included, is very similar to the land needed for PV. We have already seen that the same can be said about land for hydroelectricity. The idea that PV requires more land than we can afford is a myth. Even in populous areas—the worst-case scenario—plenty of roof space and parking lots or interstates exist that could be adapted to carry PV arrays. Utility planners know that locating PV capacity on the East Coast would be a lot easier than locating a new nuclear or coal plant. Land is not the problem; but, as we shall see, cost is.

PV Cost: Round One

Let's look at a back-of-the-envelope cost estimate of a PV system that could offset a nuclear power plant on a per kilowatt-hour basis.

In Missouri, such a PV plant would have to produce about 2.4 billion W (2.4 gigawatts) at noon on a sunny day. (This is called its peak-watt capacity, the usual measure of a PV system's size.) On a yearly basis—as stated above—it would produce the same electric output as a nuclear plant—6 billion kWh.

Today's prices for installed PV systems are about $6 to $10 per peak watt ($W_p$), although some experimental systems are going in for about $5/$W_p$. Let us use $6/$W_p$. A 2.4 billion W_p PV plant would cost $14.4 billion today (2.4 billion times $6/$W_p$) without storage. However, despite the fact that a PV plant would have no subsequent fuel costs and almost nonexistent operations and maintenance costs, $14.4 billion is way too expensive. It translates to about $0.40/kWh—compared to the $0.06–$0.10/kWh that we're accustomed to paying.

But our example makes one thing clear. *Cost* of a PV system, as measured by the cost of its output electricity, is the key parameter. When costs come down, other issues—land requirements—will not stand in the way of the successful use of PV on a large scale. Costs are crucial, so let's see where PV costs come from. When we do so, we can form a better idea of how PV can become less expensive and whether it is ever likely to meet our ambitious hopes for its ultimate role.

3 ☼ PV Costs

How much do the components of PV systems cost, and can such systems ever be expected to compete with conventional sources of electricity? Typical electricity prices in the US are about 6–11 cents/kWh. The cost of the plants making the electricity is closer to 3 cents/kWh, but the projected cost of *new* conventional plants is in the 4–6 cents/kWh range. This is also the range expected for power from cogeneration, a stiff competitor for new electricity in a climate of deregulated utilities. Industry is often able to use cogeneration as a means of producing electricity because industry frequently uses high-pressure steam as a source of process heat. This same steam can also be run through a generator to make electricity. The simultaneous production of these two energies is where the name *co*generation comes from. So to be competitive, 4–6 cents/kWh is a good long-term goal for PV.

We cannot estimate PV costs without making certain assumptions concerning the nature of a PV system and the performance and cost of its components. Some of these are easy to make (such as land costs, which are well known); others depend on projections of existing technologies or expectations of improvements that go beyond existing technologies. Our aim is to provide a realistic, yet optimistic estimate of future PV economics. We do not want to paint a rosy picture and say that PV will satisfy our energy goals immediately. We are merely predicting what future PV costs could be if we put enough effort and commitment into it. The goals are not pie-in-the-sky, but neither will they be achieved in this century. They are realistic goals for a future in which PV will be regarded as a mainstream technology.

To do justice to our purpose, we have erred on the side of *optimism* where technical progress is concerned *because that has been the history of progress to this point.* To assume *conservative* performances or costs in order to seem more judicious would do an injustice to PV, which has consistently moved well beyond expectations. For example, some of today's cell efficiencies are beyond the levels once characterized as their theoretical limit, because innovative cell designs have been developed in the interim. Meanwhile, costs of manufacturing some PV modules have dropped below levels once characterized as impossible. Again, to underestimate the potential of PV would falsely hamper the commitment needed to develop it.

All the projections assumed here are realistic in the sense that they are based on what we think are practical advances over existing technologies. For instance, although we assume a lower cost for PV support structures (i.e., trackers) than has generally been accepted, even lower costs have been hypothesized in some recent studies (e.g., one from Princeton's Center for Energy and Environment) but are not assumed here. To use the lower costs of the Princeton study would almost make PV too easy. We have not done so. It is also very likely that PV module costs will be *lower* than those we assume. This is based on the fact that the cost of the materials going into PV modules is significantly less than the cost we assume, yet in most mature industrial products cost is very similar to the sum of total materials costs. So despite the fact that we are optimistic about the future of PV, we are not even close to pushing up against theoretical limits in our assumptions. In 30 years, our cost estimates will probably seem stodgy and shortsighted. PV will be even cheaper than we are projecting here.

We will now examine the three main aspects of PV cost: PV modules; structures that support PV modules; and electronic equipment and storage options that make PV-generated electricity appropriate for general use. We will also develop an easy method of estimating PV costs in familiar units, cents/kWh (the same units as on our electricity bill). We will use the same method throughout this book to make comparisons among PV technologies. The basis of this simple technique is the division of annual cost by annual electricity production to give an average cost of PV electricity.

Past Mistakes

Past cost projections of PV have varied greatly. Some analysts of the early 1970s got into trouble by making conservative assumptions. They incorrectly assumed that PV would forever be too expensive because they made simplistic—though they probably thought, pragmatic—assumptions about PV technologies. They thought the PV technologies that existed then would always be the only choices. When they made cost projections, they just tweaked and massaged their existing numbers—taking 10% off here and 40% off there. Few had the courage to project tenfold or more cost reductions. But these major reductions have actually occurred.

The author knows this problem well because he once had a heated exchange of letters in a trade journal, *Chemical and Engineering News* (Sept. 29, 1986), with a reader who had read such analyses from the 1970s. The critic of PV told the author—quoting outdated projections—that PV would always cost too much. He said it would be a hundred to a thousand times too much. He even quoted ultimate minimum PV cost projections from a *1972* report. *In actuality, his ultimate cost had already been surpassed when he wrote his letter.* He underestimated PV and the ability of our scientific community to develop its potential.

Of course, the error can be, and has been, made the other way: overly rosy pictures of rudimentary PV systems have caused optimism for near-term options to be too great. Numerous papers have predicted the cost-effective use of PV by the next decade. Many of those "next decades" have now passed with no tidal wave of PV systems. And critical technical issues—from performance instability to the grossly high cost of manufacturing—have been dismissed as minor barriers. Some PV options have been buried beneath such "minor" barriers. They were simply incapable of reaching the efficiency, cost, and durability goals needed to make them competitive.

Back to Cost: PV Systems

The components of a PV system vary with application. In its simplest form, a PV system can be a PV panel and an external circuit to connect it to a load (e.g., a pump). In this case, the load must be able to use direct current (DC) electricity, because that is what PV panels produce. Examples of this simple system are PV battery chargers, some PV-driven water pumps, and solar calculators.

Yet even the simplest PV application may require some minimum electric storage to be effective. Storage is a ubiquitous problem for solar energy. Solar energy is totally dependent on sunlight: under cloudy conditions, output disappears or is much reduced. At night, no output is possible. But many valuable applications require continuous, uninterrupted power. Storage of PV-generated electricity is the only way to provide power on demand.

Unfortunately, storage costs money. It is a significant burden on PV costs, greatly complicates PV systems, adds maintenance demands, and the need for it makes PV harder to use. But storage is in no way an insurmountable problem. Properly done, storage can become the vehicle through which PV can meet almost all of our energy needs. Even without storage, various utility studies have shown that PV could be used for 5 to 15% of all US electricity, which is about the same level as nuclear or hydroelectric.

The basic components of a typical PV system are PV modules, support structures and possibly sun trackers, and land. We will take the following approach in our cost estimates: first, we will estimate the cost of providing DC electricity from PV. This is the kind of electricity that could subsequently be used as input to a practical PV system. Three types of systems can use this DC electricity: one that uses the DC electricity for a specific purpose (say driving a DC industrial motor); one that changes the DC electricity into alternating current (AC) electricity and feeds it into a utility grid; and another that stores some portion of the electricity and then uses it as needed, perhaps to provide *upon demand* a continuous, dependable supply of electricity.

Thus, we will analyze three kinds of PV systems:

1. DC electricity production
2. AC production with no storage
3. AC production with storage

The last of these is perhaps the most useful of the PV options because it corresponds to the public's expectation of a continuous power source that can be tapped on demand. The first—simple DC electricity—needs to be understood because it provides the building block—raw electricity—from which practical PV systems are built. The second case—intermittent AC electricity—is useful because at least one major, near-term utility application, called peak power production, can exploit this kind of electricity. Various plans are already on the drawing board for using PV in significant sizes for peak power production.

Figure 6 shows a PV system configured to supply utility power. Cells are built into modules; modules are assembled into an array and carried by a support system. The support may be fixed in place or it may be used to track the sun across the sky to maximize the amount of sunlight that strikes the array. Wiring from the array is brought to a power-conditioning apparatus (also called an inverter), which in this case must change the PV-generated DC to AC and remove spikes or fluctuations that would make it inappropriate for general use. AC is a form of electricity in which the direction of current flow reverses periodically; and it is the kind of electricity that we use in our everyday appliances. The PV-generated AC electricity is transmitted from a substation along the utility's power lines to the customer—us. This type of system is like our second example: it supplies AC power but has no storage.

A variation on this is the rooftop PV array (Figure 7). If such arrays are isolated from the utility grid, they would need storage. Right now, batteries are the only off-the-shelf choice for storage (Figure 7a). On the other hand (Figure 7b), if the PV user were already on the utility grid, they could connect the PV system to the utility's power lines. During the day, when they had more power than needed, they could sell it to the utility. At night, they could buy it back. This is like

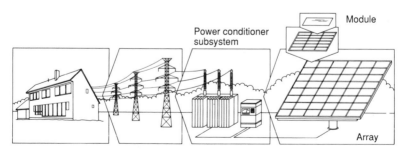

Figure 6. A PV system connected to the utility distribution grid consists of PV modules, arrays, power-conditioning, power subsystem, transmission, and the consumer.

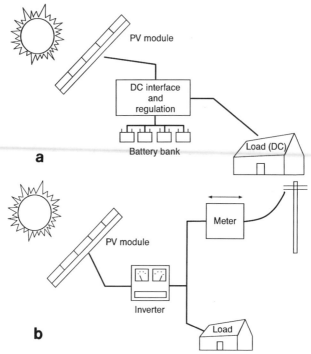

Figure 7. (a) Schematic of a stand-alone PV system with battery storage to provide dependable DC electricity day and night. (b) Even for a home connected to the utility grid, PV can be of use by producing electricity during the day. Excess PV electricity can be sold to the utility; at night or under adverse weather conditions, the homeowner can purchase electricity from the utility.

our third example, in the sense that the existing utility grid—which soaks up the excess PV power during the day—acts like an infinite, nearly 100%-efficient, storage facility.

Flat Plates and Concentrators

Before we go on, we have to define two kinds of PV modules: flat plates and concentrators. They have rather different system characteristics. Flat plates are the kind we've already discussed. They are large, flat panels assembled into even larger arrays. They can be mounted on tracking or nontracking supports, whichever makes more sense economically for the specific situation.

Concentrators are based on a rather clever idea for minimizing total cost. Figure 8 shows one. Large, concentrating lenses focus sunlight on small cells. Another approach would be to use large mirrors to concentrate sunlight on cells. In theory, large areas of lenses or mirrors can replace large areas of presumably more expensive cells, reducing total module cost. In practice, added costs from other components needed for concentrators can raise costs substantially. Both flat-plate and concentrator modules are being developed.

From a system standpoint, the difference between flat plates and concentrators occurs because the most promising concentrators demand precise, two-axis tracking to focus continuously on the sun. In addition, they can use only direct-beam sunlight, because they cannot focus diffuse sunlight. That means that the diffuse portion, at least 20% of the sunlight even in a desert, is unusable. A concentrator does not produce any energy when the weather is cloudy and the diffuse component of sunlight is very large. In contrast, flat plates work on

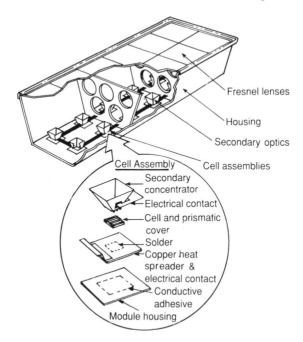

Figure 8. An advanced design showing most of the components of a concentrator module. Small, high-efficiency cells are at the focus of concentrated sunlight.

cloudy days just as they do on sunny ones, producing as electricity a fixed fraction of the available sunlight. Concentrators are geographically limited to sunny locations, while flat plates are not—which is one of their major advantages.

The potential cost of manufacturing flat-plate and concentrator modules differs. For flat plates, PV modules are just an aggregate of interconnected cells. For concentrators, a module is much more complex. In the case of concentrators with lenses, it consists of the lenses, the cells upon which the lenses focus sunlight, a boxlike structure carrying these lenses/cells, interconnections between the individual small cells, and passive cooling components such as fins on the back that must be there to dissipate excess heat. To offset the greater module cost of concentrators, the cells that go into concentrators must be much more efficient than the cells used in flat plates.

PV System Performance

To understand PV electricity cost, we must also understand how much electricity a system produces. Electricity cost is the system cost divided by the system's electrical output. The more electricity a system of a given cost produces, the lower is the unit cost of that electricity to the consumer.

PV system output depends on: (1) module efficiency during operation, (2) losses in the DC PV system and subsequently in power-conditioning and/or storage, and (3) the locally available sunlight and the tracking design.

All PV modules lose a small amount of efficiency with increased temperature. This loss varies somewhat among different PV technologies, but as a rule of thumb about 0.2% loss per each degree Fahrenheit is expected. The operating temperature of a PV module is the temperature of the cells themselves, which can be quite high (over 110° F) since they are terrific absorbers of sunlight. A PV module that is 15% efficient at 80° F might be 14% efficient at 110° F. Concentrators, which are usually located in the desert, suffer from higher ambient temperatures and higher operating temperatures (140° F).

But module efficiencies are measured under standard conditions at about 77° F (25° C). An acceptable value for efficiency losses under actual operating temperatures is 10%.

In principle, concentrators could have an even more critical problem with overheating than just the local temperature. Since cells are at the focus of concentrated sunlight—which may be 50 or 1000 times as intense as normal sunlight—they would be expected to get extremely hot. Temperatures well over 1000° C would be possible under concentrations of 1000 times normal sunlight. But cooling via heat dissipation through heavy, metal fins on the back of the modules is used to prevent this kind of overheating. Concentrators can actually operate at levels above ambient temperatures that are close to those of flat plates.

But concentrators with lenses have another problem: some light is lost in the lenses during focusing. Normally, glass lenses are quite efficient. But glass lenses are too expensive for the kind of affordable costs required to make concentrators viable. A specialized lens made of plastic, called a Fresnel lens, is used instead. These lenses are much cruder than glass lenses and have raised edges that cause losses. At present, Fresnel lenses block or reflect about 12% of the light. We will assume that modest improvement will occur and use 10% lens losses in our estimates for concentrators.

All systems—flat plates or concentrators—have minor efficiency losses from phenomena such as: dirt accumulation on the panels; shadowing by other parts of the array; wiring losses; DC substation inefficiencies; and power-conditioning inefficiencies. Dirt accumulation, self-shading, DC substation inefficiencies, and wiring losses total about 5%. DC-to-AC inverter losses can be about 3–7%. Larger losses (10–50%) can occur from electric storage but are very dependent on the specific storage option. We will return to storage and inverter losses when we get to our more complex systems.

The variation of sunlight by location, season, climate, and time of day is a major consideration. We will not choose a typical US site (such as Missouri) for estimating PV costs because the option of using PV in desert locations is a good one and should not be ignored. Picking a less sunny location would obscure that fact. Another reason is that concentrators—a key PV option—are only practical in relatively cloud-free regions, so we would not get a fair picture of their potential if we chose an average US climate.

On the other hand, for flat plates our assumption of a very sunny PV site, such as Phoenix, yields costs that are only about 50% better than they would be for even the *worst* PV site in the US, such as in Seattle or Buffalo. PV costs have come down orders of magnitude;

differences of 50% are not going to make or break PV, even on a local level. The choice of sites is not critical to the economics of flat-plate PV, where costs will only vary by about the same amount that conventional electricity costs already vary geographically, though for different reasons.

Phoenix, Arizona, is an exceptionally sunny location typical of southwestern US climates. Phoenix receives about 3.3 MWh of global sunlight each year on each square meter. The various amounts of available light for fixed, and single- and dual-axis trackers can be found for flat plates and concentrators. (Concentrators use two-axis tracking but cannot use diffuse light, so their available sunlight is lower than two-axis tracking for flat plates.) The results are given in Table 3-1.

The annual sunlight values in Table 3-1 have been measured rather than calculated. They are reasonable for typical years of available sunlight in a Phoenix-like location. Notice that in Phoenix the fixed flat plate gets 24% less sunlight than the flat plate on a two-axis tracker. One-axis tracking is right in the middle between fixed and two-axis tracking, providing about 12% less than two-axis.

Because of the loss of diffuse sunlight, a two-axis *concentrator* gets about the same amount of usable sunlight as a fixed flat plate—but then loses another 10% because of lens losses. One can get a sense of the amount of diffuse light available in even the sunniest regions (i.e., Phoenix) by comparing the amount available to the two-axis flat plate with the amount available to the concentrator: the concentrator gets 21% less sunlight (before lens losses). This loss is from the loss of diffuse light and is a *minimum*, based on the nearly cloudless conditions assumed in this example. In most other regions, concentrators would lose much more (40% or more) to cloud cover. That is why

Table 3-1. Solar Availability in Phoenix[a]

Tracker/system	Annual sunlight	Reduced by system losses[a]
Fixed flat plate	2500 kWh/m²-yr	2125 kWh/m²-yr
1-axis flat plate	2900 kWh/m²-yr	2465 kWh/m²-yr
2-axis flat plate	3300 kWh/m²-yr	2800 kWh/m²-yr
Concentrator	2600 kWh/m²-yr	1950 kWh/m²-yr

[a]For a DC system. Wiring, shadowing, DC substation, and dirt loss (5%), temperature loss (10%), and lens losses (10%), where applicable, i.e., for concentrators. No power-conditioning or storage losses.

concentrators are generally considered to be limited to very sunny locations like the desert Southwest. Flat plates, on the other hand, have almost no geographic limit. They can compete with concentrators in sunny regions and will probably outdo them in more typical ones.

Balance-of-System Costs

PV system costs can be broken down into two major categories: PV modules and so-called balance of systems (BOS)—everything but the modules. In most scenarios, the PV modules are about half the total system cost.

Some pertinent BOS categories are:

1. System design costs
2. Land area upon which to build the PV array
3. Site preparation, roads, fences
4. System installation
5. Two-axis, one-axis, or fixed support structures and their foundations
6. Power-conditioning equipment and related electrical system costs
7. Operation and maintenance
8. Indirect costs (contingencies)
9. Fixed costs (interest expenses for capital)
10. Storage and related costs, where applicable

Analyses of these BOS costs are often broken down into area-related and power-related costs. That is, some (like installation, support structures, and land) are intimately linked to the *area* of the array rather than its output. The more module area needed to meet an electric load, the more support structures and land area are needed for the modules. Others, like power conditioning and storage, do not depend on area *per se*; they depend on total power. BOS costs by area are estimated in units of $/m^2$, where "m^2" refers to module area. BOS costs, which are enumerated by $ per installed peak kilowatt, are estimated in units of $/kW_p$ of module peak power.

To get *total BOS cost*, one multiplies the area costs by the total module area; multiplies the power-related costs by the total peak power; and then adds these two together. All of this may seem rather complex. But remember, we are trying to discover whether a new, clean system of producing our electricity can compete with nuclear and fossil-fuel electricity. It is worth the effort to define the costing method well enough to get a sense of its credibility.

The following are some comments on BOS costs by category:

1. *System design costs* are generally considered minimal. For example, the largest PV demonstration project to date, the ARCO Solar Inc. 6-MW plant at Carrisa Plains, California, took less than a year to design and install. Fairly extensive literature on the design of PV systems exists and can be drawn upon. Presumably in future systems—the ones that we are interested in from the standpoint of PV's potential—these costs will be even smaller. We may think—based on all this talk of different kinds of trackers, different PV modules, etc.— that PV designs are complex. Actually, compared to conventional generators, they are a piece of cake. Just think of all the fuel handling, pollution control, and safety equipment that aren't needed. Also, PV systems are modular, in the sense that designing a 10-*billion*-watt system is very much like designing a 10-*million*-watt system. Bigger systems are just many smaller ones put together. For a reasonably sized system, design costs are likely to be under $1/m^2$.

2. *Land area* is a small fraction of BOS costs. Many people find this hard to believe, but it is an incontrovertible fact. For instance, *prime central Illinois farmland* goes for about $2000 per acre. There are 4400 m^2 in an acre, which means each square meter goes for about 50 cents. Since there are about 2 m^2 of land for every square meter of array, we must double our land cost and use about $1/m^2$ in our example. Even this is insignificant. As we'll soon see, affordable BOS costs are in the $30–$70/m^2$ range, almost two orders of magnitude more than land costs.

Based on $1/m^2$, we could afford a lot of prime agricultural land for PV if we needed it. But we probably won't unless farmers choose to use it that way because it enhances their income. We will probably end up using a lot of useless scrubland in the Southwest and rooftops in the Northeast, and these will cost much less. But the point is—made this time in the most obvious, bottom-line way—*land in the US is not a problem* either in terms of cost or in terms of the total area needed to provide all of our energy. The only problem with land or array area is the fact that in some locations there may not be enough available space right where it is desired. Even this is probably rarer than it might seem, except within cities. Land costs themselves are such a small part of PV costs that getting land *free* makes little impact on overall PV costs or profitability.

Generally speaking, land requirements are up to three times larger for two-axis trackers than for single-axis or fixed arrays. This is because of the way that trackers move and potentially shadow each other when the sun is low in the sky. Land costs could be an issue for two-axis trackers (e.g., for concentrators) if the most expensive land were chosen; but that is an improbable scenario, since concentrators are likely to be used in waterless, desert locations—not much use otherwise and not highly priced.

3. *Site preparation* consists of clearing away obstacles, grading the land under the arrays to some reasonable flatness, and providing access roads. In an EPRI study (Electric Power Research Institute; an arm of the electric utility industry), site preparation was allocated a fairly large cost. Costs for concentrators were twice those of flat plates, because of the need for preparing the ground before installing precise two-axis sun trackers. In most studies subsequent to EPRI's, site preparation costs were regarded as smaller. We follow those studies and are assuming about 5% of area-related BOS (i.e., about $5000/acre, or $2/m^2).

4. *Installation costs* have been regarded as significant. Installation includes setting up the modules and arrays and wiring them together. (Support structure costs are included in another category.) In one study by EPRI, installation labor costs were almost 15% of area-related BOS for flat plates and 20% for concentrators. In that study, large costs were associated with the actual construction of arrays in the field. But PV manufacturers are now considering a different approach in which modules are attached to array supports at the factory and then shipped on flatbed railroad cars to the installation site. Factory installation is always cheaper than field installation. Installation costs could be quite a bit less than those of EPRI studies. We will assume that installation costs are about $5/m^2, which is about 10% of area-related BOS.

System installation costs will be slightly greater for concentrators than for flat plates due to the need of concentrators to have precise tracking. Two-axis trackers are more expensive to install than fixed arrays or one-axis arrays. Also, flat-plate arrays should be less cumbersome and easier to transport from the factory without breakage. But because such variations will have a minor overall impact (and for the sake of simplicity), we will use the same approximate figure ($5/m^2) for all of them.

Now we get to some *real* cost-drivers:

5. *Support structures and trackers* are *significant* BOS costs (about 75% of area-related costs). Generally, fixed supports are the least expensive and the most rugged; single-axis trackers can be nearly as rugged and somewhat more expensive; two-axis trackers can be expensive and require more maintenance, especially those for concentrators.

The cost of supports/trackers and the foundations on which they are built are major BOS expenses. *Present* costs are about:

☐ Fixed structure, $55/m^2$
☐ Single-axis tracker, $85/m^2$
☐ Two-axis tracker, $120/m^2$
☐ Two-axis tracker for a concentrator, $140/m^2$

An important Department of Energy (DOE) report, the Five Year PV Research Plan 1987–1991, assumes that these same structures will someday reach the following costs:

☐ Fixed structure, $35/m^2$
☐ Single-axis tracker, $55/m^2$
☐ Two-axis tracker, $90/m^2$
☐ Two-axis tracker for concentrator, $110/m^2$

Consider these costs alongside the totals for all the previous categories, i.e., about $9/m^2$. With the cost of the PV module, the cost of the support structure/tracker is the major PV cost-driver.

6. *Power-conditioning* is another major BOS cost, although it is significantly less than the support structure cost. By power-conditioning, we mean everything needed to change a PV system's DC current to AC; to smooth it so it can be used by the most demanding utility customer; and to regulate it and transmit it along the utility grid. The main costs are a DC subsystem (the part of the electrical system needed to make the array output smooth), a transformer, the DC-to-

AC inverter, and the AC subsystem needed to connect the system to an external grid. Of these, the inverter is the most expensive; but its cost is likely to fall the fastest as demand for more of them for PV increases. These costs are put in terms of dollars per kilowatt-hour of peak installed capacity because they scale with system power, i.e., the more power an array produces, the more capacity is required of the power-conditioning equipment. In an EPRI study, power-conditioning costs were about a fourth of the other BOS costs. Sandia National Laboratories estimates them to be $200/kW and $100/kW in 1986 and in 1990, respectively. In the DOE analysis, they were about $150/kW.

Power-conditioning costs are about 20–40% of the other array costs. We will assume the Sandia estimate ($100/kW) because existing estimates are based on buying inverters that are almost one-of-a-kind. Once PV begins to make an impact, better solid-state DC-to-AC inverters should become available. When we make various cost projections, we will assume that half the cost ($50/kW) will be attributable to the DC portion of the system, and half to the inverter and the AC system.

7. One of the expected strengths of PV systems should be very low *operation and maintenance (O&M) costs*. Recently the president of ARCO Solar, Dr. Charles Gay, characterized the O&M of their 6-MW Carrisa Plains PV plant as essentially zero. Obviously PV plants need no fuel. But just as important in these days of high maintenance cost is that PV can produce power without much supervision or repair. Two of ARCO Solar's major PV demonstrations, at Carrisa Plains and Lugo, California, have been operated *solely by computer with no on-site personnel* for several years.

The only O&M of PV plants should be maintenance to fix any breakdowns associated with sun trackers. For fixed or one-axis trackers, these costs could be negligible. Two-axis trackers used for concentrators may be more expensive. In general, recent studies suggest a range of O&M cost of about 0.02–0.1 cent/kWh—much lower than for conventional technologies. These costs are almost too small to consider in a cost estimate of PV.

8. *Indirect costs*, or contingency costs, were assumed to be about 50% by both EPRI and DOE. By indirect costs, we mean a factor that is used to multiply all other costs to take into account such things as what EPRI called costs for "engineering fees, contingency, owner's cost, marketing and distribution, and interest during construction."

For simplicity, this can be thought of as an overhead rate. Since it multiplies all other costs significantly, it is a major factor. In a recent paper, Sandia National Laboratories used 25% instead of 50% as the indirect cost for their estimates. They said the choice was "based on current system procurement experience," i.e., Sandia had actually procured PV systems at the lower indirect rate. They characterized the DOE/EPRI estimates as based on the contingency costs of other, conventional plants in which there were up-front costs for having spares to replace large equipment such as gas turbines and nuclear reactors. PV is both more reliable and more modular than conventional generating technologies: contingencies will consist of a few extra modules. Current failure rates for modules in most systems are as low as 0.01% (1/10,000), suggesting that contingencies for PV will be very minor. So it seems that this very significant cost (in that it expands all costs by a factor of either 50% or 25%, depending on which is the assumed rate) will be lower than the DOE and EPRI expectations.

9. Some interest rate is assumed to cover the loan secured to finance the PV system. (This is often called the fixed charge rate.) It can be used as a factor to multiply the total cost to get an annual cost. For instance, if one has a 30-year mortgage loan at 10% interest, one's annual payment is very close to 10% of the loan per year. (The actual fixed charge rate associated with a 10% thirty-year loan is slightly higher, 10.53%, which accounts for the extra amount needed to pay back the principal of the loan over 30 years.) The product of the annual loan rate and the indirect rate is the factor we multiply our total system cost by to get an annual cost. The fixed charge rate assumed by EPRI and DOE in their studies is 9.1% based on existing utility estimates of their loan rates.

10. Storage. By storage, we mean a rather complicated system depending on some means of storing DC electricity. Examples of storage are: batteries; hydrogen production and conversion back to electricity; pumped hydroelectric; and superconductors. Each has its own typical costs and efficiencies. As a rule of thumb, storage system might double the cost of electricity versus that coming from an intermittent PV system without storage. Because storage costs are technology-specific, we will not go into more detail until we examine one option, hydrogen production via PV, at the end of this chapter. Also, see Chapter 15 for an extensive discussion of storage.

Comments on Area-Related Costs

For an intermittent DC PV system, almost all BOS is area-related. A recent (1987) Sandia National Laboratory report separates them into tracking types (this being the largest component) and gives the *present* area-related costs as:

- □ For a fixed structure, $85/m^2$
- □ For a single-axis tracker, $135/m^2$
- □ For a two-axis tracker, $200/m^2$

These include everything—land, wiring, installation, site preparation, and structures. Sandia is a DOE laboratory with major input in setting DOE goals and standards for PV. Sandia suggests that area-related costs should be $50/m^2$, $77/m^2$, and $124/m^2$ for nontracking, single- and two-axis trackers, respectively, in the 1990s. For very precise two-axis tracking for concentrators, they project a slightly higher cost, $140/m^2$.

If we added together our estimates for a fixed flat-plate system, we would get about $44/m^2$: $35/m^2$ for the support structure plus $9/m^2$ for everything else. This is $6/m^2$ less than the Sandia projection.

Sandia's projected area-related costs for the simplest of PV structures, fixed flat-plate arrays, can be considered pessimistic. The only significant structural element in a nontracking support is a tilted, flat surface able to withstand wind loads. In a recent report prepared by the Princeton Center for Energy and Environmental Studies, about $20/m^2$ was assumed for the support structure and $33/m^2$ was assumed for area-related BOS (instead of $50/m^2$). This came about after discussions with PV manufacturers and examination of some existing systems that already approached the Sandia cost of $50/m^2$.

We are going to use the Sandia figure of $50/m^2$. We will *not* use the Princeton figure because no consensus on its applicability has been developed. Similarly, we will assume that our own figure of $44/m^2$ is low. But we remind the reader that all costs found here could be reduced by about 0.5–1 cent/kWh by assuming one of the lower figures.

The same Princeton study pointed out other estimates of even lower area-related costs—as low as $10/m^2$. These extremely low costs were based on very innovative/controversial designs. In addition, there are recent reports of very low costs for two-axis trackers—as low as $60/m^2 (instead of the $124/m^2 assumed by DOE). Trackers can get about 30% more sunlight than nontrackers, which would reduce all costs by the same figure. (Note that we are assuming about the same cost, $50/m^2, for a fixed array.) Although we believe that costs below $50/m^2 for fixed arrays and below $124/m^2 for trackers will eventually be approached, we will not assume them here because they would literally make PV too inexpensive. However, the potential for even lower costs than those found here (by significant factors of about 30–60%) should be kept in mind.

Putting it Together

We want to understand how much PV costs and what makes it cost that much. To do so, we will use three kinds of estimates: one for the simple case of providing intermittent, DC electricity; one for intermittent *AC* electricity; and one for continuous AC electricity based on storage. We will start with the simple case of providing DC electricity. This is a kind of feedstock electricity, since it can be a building block out of which more complex and useful electricity is developed. In some cases, such as an industrial setting, where the use of large amounts of DC electricity is common (to power DC motors), intermittent DC PV could be a practical option.

We have estimates of BOS costs and sunlight available for PV electricity production (Table 3-1). We need two more things to proceed: the *cost* and sunlight-to-electricity *efficiency* of our PV *module*. By adding the module cost to the BOS costs, we could get our total system cost. Then we could use the module efficiency to calculate how much electricity our system produces.

But instead of assuming a PV module cost and efficiency, we can turn the whole problem around. We can decide upon a goal for PV electricity that would make it competitive with conventional electricity production. Then we can calculate how much we could afford to pay for our assumed module and still reach that cost goal. Suppose we desire to have DC electricity costing 4 cents/kWh. What can we afford to pay for a module of a given efficiency? Is such a cost a practical goal? Answering these questions will give us an idea of whether we can ever be competitive with conventional power sources such as nuclear, coal, oil, and natural gas. Our method is simple: we will add up all the costs (some unknown module cost and BOS without DC-to-AC inverter or storage), divide them by the annual DC electricity production, and then multiply the whole thing by an annual fixed charge rate. Setting this equal to 4 cents/kWh, we can solve for the module cost. Then we'll decide if our assumed module efficiency and this cost are reasonable goals for PV modules.

DC Electricity: A Simple Example of Cost/Performance Analysis

Let's assume that we will someday have 15%-efficient flat-plate modules. (Achieving this efficiency for a very low-cost technology is not the case now but is consistent with DOE research goals.) We want to know how much money we can afford to pay for our modules and still produce DC electricity that costs us a reasonable amount, say 4 cents/kWh.

We want to use our assumptions (as above) for future BOS costs, not BOS costs as they are now, because we want to know what our long-term module cost goals should be. (We already know that our present costs for modules are too high, otherwise PV would be cost-competitive now.) Taking the relevant cost assumptions from our earlier comments:

☐ Area-related costs for a fixed flat-plate structure: $50/m^2
☐ Power-related costs of $50/kW$_p$ (no inverter; we want DC, not AC costs)
☐ Indirect rate of 25%
☐ 9.1% fixed charge rate

The area-related costs ($50/m^2) are our total cost except for the cost of the DC subsystem and the module, the latter of which is what we want to determine. Our power-related costs can be translated into a cost per square meter by multiplying the power output of a square meter of modules (0.15 kW) by $50/kW, to get about $7/m^2.

Using the Sandia estimate of indirect costs, 25%, and a fixed charge of 9.1%, one multiplies all costs by 11.4% (0.091×1.25) to get an *annual* cost per square meter.

Table 3-1 says that a fixed flat plate in Phoenix receives about 2125 kWh/m^2 annually after losses for wiring and operating temperature are subtracted. At 15% efficiency, about 319 kWh/m^2 of DC electricity would be produced by the system.

Thus, the cost of DC electricity on a per kilowatt basis would be the area-related BOS ($\$50$/m^2) plus the cost of the power-conditioner ($\$7$/m^2), plus the cost of the module (the variable we are solving for) multiplied by the annual fixed charge and the indirect rate (11.4%), all divided by the annual system electricity output (319 kWh/m^2):

$$(\$50 + \$7 + \text{module cost in } \$/m^2) \times .114/(319 \text{ kWh/m}^2)$$

This is really only the annualized cost divided by the annual energy production, which naturally gives a cost in $/kWh. However, in our example, we are trying to find out what our allowable module cost is if we expect to have 4 cents/kWh DC electricity. If we set our equation equal to $0.04, we can solve to get an allowable module cost of about $55/m^2. This is the allowable cost of our module if we expect to provide DC electricity costing 4 cents/kWh. Our module cost ($55/m^2) is within $5/m^2 of our assumed area-related BOS costs, giving us a sense of their similar relative magnitudes in this example.

The Peak Power Example

The next case is the next simplest: we simply add a power-conditioner to the system and sell the intermittent AC electricity to a utility. Is this a realistic application?

The DOE chose this as their typical system for their Five Year Research Plan 1987–1991. The intermittent AC system that we describe actually corresponds to one that utilities would find useful to provide electricity for their peak summertime air-conditioning demands—a fairly large (many billions of watts annually) and rapidly growing market. PV would be assigned a high value for this example if its production matched the utility's peak need. In many cases, such as in southern California where air-conditioning dominates electric demand, PV is a perfect match for peak demand (Figure 9). We will find later that this market—providing peak power—is one that PV manufacturers consider to be their most lucrative in the near term (before 2000 AD).

To adapt our equation to this second case, we need to do two things: reduce the PV system efficiency by 0.97 to take into account losses in the inverter and add the cost of an inverter to our other BOS costs. Reducing the system efficiency by 0.97 means that we can assume that only 309 kWh are produced annually by a fixed flat-plate array in Phoenix. We have assumed an added inverter cost of $50/kW. Multiplied by the peak power of our system, 150 W/m^2, we get about $7/$m^2$ for our inverter. This changes the BOS costs from those assumed above ($57/$m^2$) to $64/$m^2$. Let us assume that we are using the same modules as in our previous example, i.e., those costing $55/$m^2$. We can find the cost of our AC electricity by using our equation and substituting our new BOS costs (with inverter) and our slightly reduced annual electrical output:

$$(\$64 \text{ BOS w/inverter} + \$55/m^2 \text{ modules})$$
$$\times .114/(309 \text{ kWh}/m^2) = 4.4 \text{ cents/kWh}_{AC}$$

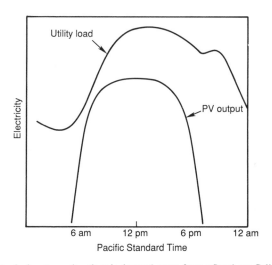

Figure 9. A typical summer-day electric demand curve from a Southern California utility and the daily output of a PV array. Because PV output is a good match for demand, its value is greatly enhanced.

So, our output cost of AC electricity (no storage) is 4.4 cents/ kWh$_{AC}$, about half a cent more than the cost of intermittent DC electricity. This figure, 4.4 cents/kWh, is the projected cost of our intermittent AC electricity being supplied to a utility for its peak power requirements. Since peak power requirements can be a utility's most expensive to supply—with prices as high as 15–20 cents/kWh in existing systems—4.4 cents/kWh would be a very attractive cost for PV. This would be for a Phoenix site. But PV would remain attractive for a Midwest or Atlantic seaboard site. By choosing a Phoenix-like site, PV costs are only 50% better than the absolutely worst sites in the US and only 25% better than average sites. This means that for typical US sites—including the Midwest and even the Atlantic seaboard—AC PV for peak power without storage (using our assumptions) would cost under 6 cents/kWh, a very competitive figure.

Continuous Power from PV

Perhaps the most interesting and ultimately useful example is the one in which we assume some sort of storage. By using storage, PV could provide controllable amounts of electricity under cloudy conditions or at night. PV could potentially be used for everything that conventional electricity is used for if the economics of this example were favorable.

For an example of storage, we have chosen the on-site production of hydrogen using PV-generated electricity. Although other methods may eventually supersede it, this is an easily understood system with much flexibility. It can be used to provide DC electricity, AC electricity, or even hydrogen as a fuel. For the most part, it does not assume any sort of breakthrough outside of the PV assumptions. For much of this example, we will follow the method and assumptions of a report by Joan Ogden and Robert Williams of Princeton University's Center for Energy and Environmental Studies.

Hydrogen can be produced by an electrolyzer which splits water (H_2O) into hydrogen gas and oxygen (Figure 10). The electrolyzer efficiency assumed by Ogden and Williams was 84%. The coupling efficiency of a PV array with an electrolyzer was about 94%, which has been demonstrated in a DOE PV/hydrogen system at Brookhaven Laboratories on Long Island. Because electrolyzers use DC electric-

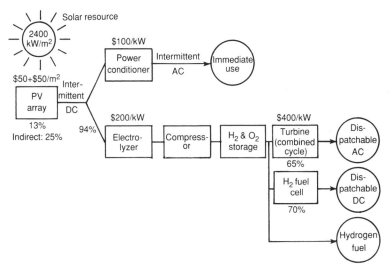

Figure 10. Photovoltaic electricity can be used to split water (H_2O) in an electrolyzer into hydrogen and oxygen. Hydrogen can then be used as a fuel (bottom), or in a fuel cell for DC electricity, or in a gas turbine to provide AC electricity. Hydrogen production by PV can smooth or shift PV output and would give a PV system the flexibility to meet various loads.

ity, the need for an intermediate power-conditioner is eliminated, saving that cost and efficiency loss (3%) of going to AC with an inverter. The hydrogen from the electrolyzer can be compressed and stored on-site for later conversion back to DC or to AC electricity.

For DC electricity, fuel cells would be used at efficiencies over 70% (perhaps as high as 90%). DC electricity is of great value for industrial purposes. Most motors run on DC, and industry actually has to use rectifiers to change the utility's AC to DC. On-site PV production of DC electricity and hydrogen (with fuel cells to convert the hydrogen back to DC) is a promising area that may be similar in scope to the peak power market. It has also been suggested that fuel cells run in reverse can be used as electrolyzers (by driving the process the other way—electricity to hydrogen/oxygen). Thus, fuel cells could be used as electrolyzers when PV energy must be stored; and as hydrogen-to-electricity sources when stored hydrogen is being used as fuel.

In a system where AC electricity is desired, various systems are possible, depending on their relative economics. Fuel cells could be used with a DC-to-AC inverter, or turbines could be used that are similar to those for burning natural gas. These produce AC electricity without need for an inverter. Ogden and Williams used the latter system. Turbine efficiency was assumed to be 50% for hydrogen burned in air and 65% if the hydrogen were burned with the oxygen saved from the electrolysis step. This is much higher than conventional turbines (which are about 33% efficient) because of several advantages of hydrogen-burning turbines. Hydrogen burning with oxygen produces superheated steam (3100 kelvin). In a conventional turbine, combustion gases heat water via a heat exchanger and produce outlet temperatures of only about 2480 K. Second, the hydrogen turbine can be a combined-cycle turbine, in the sense that the exhaust gases—water vapor—can be run through a conventional steam turbine to produce more electricity. Additional efficiencies accrue to the hydrogen–oxygen turbine by burning oxygen rather than air. Under such circumstances, heat losses due to the presence of other gases (e.g., nitrogen) are avoided, as are some unwanted combustion products (NO_x). Each of these effects can substantially increase efficiency over conventional gas turbines.

If we assume 65% efficiency for the turbine, the total round-trip efficiency of the system from DC PV to continuous AC PV is only about 50% ($0.93 \times 0.84 \times 0.65$). PV/hydrogen with turbines is not a best case for storing electricity. Hydrogen with fuel cells could be over 65% efficient; batteries could be over 70% efficient; pumped hydro at least 70%; superconductivity...eventually as high as 97–100%.

A capital cost of $400/kW was assumed by Ogden and Williams for the turbines, which is the same as for existing gas turbines. Electrolyzer cost was assumed to be about $170/kW, and storage and compression costs were relatively minor. Power-conditioning costs for the AC electricity coming from the turbines were also just like those in conventional thermal systems, i.e., were negligible because no inverter was needed.

Hydrogen production stores energy. We do not know in advance how much storage will be needed; it will vary, depending on application and location. For a cost estimate, let's choose a baseload system, i.e., one capable of some level of production year-round, day and night. This was the Princeton choice, and they purposely overestimated the amount of hydrogen production in their system (75% of the total) in order to get a conservative (high) estimate of PV/hydrogen costs. Hydrogen production smaller than their's would lower their cost estimate toward that of intermittent peak AC power, i.e., toward about 4.4 cents/kWh.

Using our assumption of 4 cents/kWh DC electricity, we can estimate the cost of controllable, AC electricity from a PV/H$_2$ system as a little over 7 cents/kWh. We can see, based on the Princeton analysis, that hydrogen production and burning adds about 75% to the cost of the input DC electricity. For a less sunny location than Phoenix, e.g., New York, total AC cost would be about 9 cents/kWh. This would be the cost of using PV for baseload electricity generation in the Northeast (using our assumptions), and it would be the amount one would use to judge PV in comparison with other baseload generators such as nuclear or coal plants.

Where Are We Now...How Much Further Do We Have to Go?

How far are we from 4 cents/kWh DC electricity? And 7–9 cents/kWh baseload PV?

Let's summarize the difference between the current state of affairs and the cost and efficiency assumptions we have made to get to the low-cost electricity of the above examples (Phoenix-like location; fixed flat plate):

	Now	Needed
Module cost	$500/m²	$55/m²
Area-related BOS	$135/m²	$50/m²
Power-conditioning	$200/kW ($20/m²)	$100/kW ($14/m²)
Module efficiency[a]	10–15%	15%
Cost of DC electricity[b]	$0.37/kWh	$0.04/kWh
AC cost w/storage	**$0.68/kWh**	**$0.07/kWh**

[a]Unreduced for wiring and dirt losses (5%) and module operation at higher temperatures (10%).
[b]No storage or inverter.

We have made certain simplifying assumptions about the module technology in this table. There are actually many different combinations of current and future module costs and efficiencies. Some modules—concentrators, for example—are aiming at higher efficiencies with higher costs. The module costs and efficiencies assumed here are just a reasonable compromise.

Summary and Conclusions

We have done the basic work of understanding how electricity cost is arrived at by dividing system cost by electric output and multiplying by an annual fixed charge. We use this same equation later when we compare different module technologies. The basic equation used above is:

$$(\text{module cost } \$/m^2 + \text{BOS } \$/m^2 + \text{BOS } \$/kW) \times .114/$$
$$(\text{system efficiency} \times \text{annual sunlight from Table 3-1})$$
$$= \text{cost of electricity in } \$/kWh$$

When using this equation, some care is needed to associate the right BOS costs with the appropriate system efficiencies. Losses associated with making DC PV (wiring and dirt losses, temperature corrections, and lens losses for concentrators) are already in Table 3-1, but losses for DC–AC inverters or storage are not because they can vary on a case-by-case basis: remember to reduce system output by losses in the inverter (about 3–7%) if AC electricity is the desired output.

What observations can we make?

1. If PV module efficiencies and costs can eventually be as favorable as we assume, PV can become a major global source of economical electricity. Even on a purely financial basis (excluding the environmental benefits of PV), PV could compete with new conventional plants. At a minimum, fear of some catastrophic rise in electricity costs in the next century would be eased.

2. System costs are dominated by storage costs and costs for PV modules, followed by trackers and power-conditioners. Ultimately, PV module costs must be brought to nearly the same level as future BOS costs. This effort—*a tenfold reduction from present module costs*, which are nearly $500/m²—*is the most critical research task for making PV practical*. In fact, for this reason, research on module efficiencies and costs has always been the crux of all PV research efforts. The simple cost analysis that we have just done shows why: the efficiency and cost of PV modules drive up the cost of PV systems. Past successes, which have reduced module costs from $5000/m² to $500/m² and raised efficiencies from 5% to over 10%, are indications that the trend is moving in the right direction.

The next chapter will be devoted to gaining a better understanding of PV devices so that we can begin to appreciate their evolution toward higher efficiencies and lower costs. The impact of bringing PV costs and efficiencies into line with our assumptions would be the development of a major new form of safer and cleaner electricity generation.

4 ☼ How PV Devices Work
Round Two

In principle, we already know how PV devices work. Light is absorbed within the device. Electrons are freed to move around. An electric field pushes the free electrons to the other side of the PV device and prevents their return. The electrons, seeking charge neutrality, flow through an external circuit, performing work. This continues as long as the PV device is exposed to light.

But to understand the true potential of PV, we need a better grasp of how they work and what can be done to improve them. Here, we will learn how the key element of a solar cell—its electric field—is formed. This will help us gain some insight into the ideal sunlight-to-electricity conversion efficiencies to be expected of various cell materials.

The most important aspect of a solar cell is the electric field that drives the cell's electric current. This electric field exists at the interface between two dissimilar semiconductors. Being able to make PV devices so that the electric field is as strong and homogeneous as possible is a key aspect of making excellent cells.

The electric field in the middle of a solar cell is fixed in place. It is there when light shines on the cell; it is also there when there is no light on the cell. Over the entire life of the cell—possibly 30 years or longer—the electric field remains unchanged. It is as much a part of the cell as are grids or semiconductors.

The electric field is present because in many cases, an electric field is naturally induced when two dissimilar semiconductors are touching. How does this happen?

First Some Semiconductor Basics

Most semiconductors have a very well-ordered atomic structure called a crystal lattice. In their most perfect form, such ordered semiconductors are said to be single crystals. Single crystals can be imagined to be an infinite repetition of a simple building block. Imagine for instance a microscopic, cubic room with atoms of the same element at each corner of the room. Then imagine this built into a huge skyscraper, millions of rooms high and wide, with every atom in exactly the same place on every floor of the skyscraper. Silicon atoms form lattices like this, although in a somewhat more complex shape than a cube.

In PV, we often deal with multielement crystals—compound semiconductors made of more than one element. An example is cadmium telluride (CdTe), which is made of equal amounts of cadmium and tellurium. The building made of these elements might consist of cadmium atoms on the floor and tellurium atoms on the ceiling, alternating as the skyscraper ascended. Again, this is not the exact shape of a CdTe lattice; it merely illustrates the nature of a multielement crystal lattice.

Many, more-complex shapes are also possible. We may have three atoms—such as in the important PV semiconductor, copper indium diselenide ($CuInSe_2$)—in which floors or ceilings of copper and then indium might alternate with a layer of selenium (copper/selenium/indium/selenium/copper/selenium. . .). The skyscraper would then be assembled from these two-story building blocks.

Because its atoms are so well ordered, a single-crystal structure is the most perfect form of a semiconductor. Unfortunately, making single crystals can sometimes be too costly to be of practical use in PV, where costs are a very important element. On the other hand, many of the properties of single crystals are maintained in materials that are less than perfect and cost much less to make.

Within ordered semiconductors, electrons have very well-defined roles. Each atom can have numerous electrons. Silicon has 14 electrons around its nucleus. Ten of these are tightly held by the nucleus of silicon and do not play much of a role in the material's chemistry. They are said to be in a completed inner shell of electrons. But the outer four electrons of silicon have a key role in holding the silicon lattice together and in silicon's electronic properties. These outer electrons are called the outer shell electrons or the valence electrons.

In general, an element's incomplete outer shell of electrons plays the most important role in its chemistry. In fact, these outer electrons are the basis of the chemical bonding that holds lattices together. Atoms are most stable when they have a complete outer shell of electrons. Each outer shell is complete when it has eight electrons. The so-called Noble gases (e.g., helium, neon, argon) are examples of stable, nonreactive elements with complete outer electronic shells. An element's chemical properties are very different when its outer shell is incomplete. Two or more atoms with incomplete shells can actually stick to each other despite the fact that they are individually electrically neutral. They stick, or bond together, if this action will enable them to complete their outer shells. This is how an aggregate of neutral atoms can form into a large, strong crystal.

Silicon is a good example. Each silicon atom has four outer electrons and needs four more to complete its outer shell (i.e., to reach eight outer electrons). Each silicon atom bonds in a crystal lattice with four other silicon atoms and shares one outer electron with each. By adding four more electrons to its outer shell, its outer shell is filled. Each of the other silicon atoms has four neighbors to share electrons with as well, completing their shells, too. The tightly bound silicon atoms become a rock-hard substance despite the fact that individually they are neutral atoms.

Besides their active role in bonding, the same shared outer electrons are the key to the electronic behavior of PV devices.

n- and p-Type Conductivity

Again for simplicity, let's continue to consider a single-crystal lattice made of one element. Let us suppose that each atom is held to four others by shared outer electrons, as is the case with silicon.

This would be a very perfect structure from the standpoint of mechanical strength and electronic stability. For instance, consider the fate of one added electron. Suppose for some reason an extra electron was wandering through our perfect lattice. Where would it be able to stop? Nowhere. It would just wander around, maybe crashing into an atom once in a while. If we put a force on it—say an electric field—we could drive it through the lattice.

Now let's inch our perfect lattice a little closer to the real world. The real world has heat. Room temperature (70° F) is 273 degrees above absolute zero Kelvin. Heat jostles things around. It does the same thing in our crystal structure. The shared outer electrons of our lattice are well bound to each other, but each has a small probability of picking up enough heat to break loose. Near absolute zero, there is little chance they'll accumulate enough heat to do so. Few would break loose. But at higher temperatures, there is a higher probability of an electron gaining enough extra vibrational energy to slip free. At some even higher temperature (that varies by bond strength), there is enough free heat energy jostling the outer electrons to make most of them break free. At this point, the solid melts and loses its lattice structure.

But our interest is in the behavior of these materials near room temperature. At these temperatures, a small fraction of electrons can accidentally gain enough heat energy to break free. Let's say that one in a trillion do so. From their standpoint, they are like our free electron wandering through a perfectly inhospitable lattice. They almost never see an empty bond like the one they came from, so they have no place to stop. They are called conduction band electrons. The place in the lattice from which they emerged is called a hole—the absence of a needed electron. These holes actually behave like antielectrons in the sense that they move freely like electrons do, except in a reverse direction when they are under the influence of an electric field. Each electron (and hole) can move freely for a very long time without encountering a hole (electron) with which to recombine. This period of free movement is called a free-carrier lifetime. While it is free to move around, the electron (or hole) is called a free charge carrier, since it can be moved by an electric field and be part of an electric current.

Our perfect, room-temperature crystal reaches an equilibrium when just as many new electrons are being freed as old ones are finding holes. This equilibrium has a characteristic, temperature-dependent number of free conduction band electrons.

This number of free electrons (and holes left behind by the electrons) created by heat within a perfect crystal is a very small fraction (one trillionth or so) of the total number of bound outer electrons. A material with this few free carriers is not much different from an insulator like rubber or wood; it won't carry much electric current because it hasn't the large number of free carriers needed to do the job. But the real world has another quality that helps us make good devices from semiconductors: impurities—atoms that are not of the same kind as in the pure lattice. Normally one would think that impurity atoms would *harm* our devices. Actually, the *controlled* incorporation of impurities is the key step in making them work.

Impurities

Impurity atoms can be used in semiconductors to alter their electronic properties in *favorable* ways. The critical quality of these atoms is the number of outer electrons they have. Suppose our crystal lattice is made up of atoms (like silicon) with four outer electrons. The kind of impurities we are interested in are those with one electron more or less than this: those with five outer electrons like phosphorus; or those with three, like boron. We will find that we can do powerful things by introducing such atoms into our heretofore perfect crystal lattice.

Suppose we add one of these impurity atoms with five outer electrons to our silicon-like lattice. In practice, this can be accomplished by adding phosphorus to a silicon melt and then solidifying it. The impurity atom will (rather conveniently) mimic the behavior of our silicon atoms and sit comfortably in the lattice in place of a silicon atom. Four of its five electrons will take part in the bonding that holds the lattice together. However, its fifth electron will be extra, a third (fifth?) wheel. It will not take part in bonding; it isn't needed to strengthen the lattice.

The excess atom isn't totally free to wander off, either, because the nucleus of the impurity atom has one more proton to hold it in place. However, this force holding the fifth electron in place is much *less* than the force holding the other outer electrons in place, because those are part of bonds. At any given temperature, the probability of the extra electron breaking free is quite high. Perhaps 90% of these fifth-wheel electrons break loose at room temperature. How many is that? We may have substituted only one of these impurities for a million regular atoms; but if 90% of them are free at room temperature, there will be a million times more free electrons in the conduction band than there had been before. Their presence changes the electrical behavior of the semiconductor radically.

Such a semiconductor (one with extra electrons) is called n-type (negative-type) because it is dominated by numerous free negative charges, electrons. The impurity atoms that are introduced to add the extra electrons are called donors, because they end up donating a free electron.

p-type (positive-type) semiconductors also exist; they are dominated by holes—positive free charges. p-type semiconductors are created when impurities with one *fewer* outer electron are introduced in the lattice. These atoms also sit in lattice sites as regular atoms. Because they have one fewer outer electron, one of their bonds with a neighbor is missing a desired electron. The bond would be stronger if the added electron were present, even though the impurity atom does not need it for charge neutrality (it has one fewer proton in its nucleus than does silicon).

In this case, there is a good chance, even at room temperature, that an electron from a neighboring bond will accidentally gain enough thermal energy to hop into the bond with a hole in it. Quantum mechanics works in strange ways. The amount of energy needed for an electron to hop into a nearby hole is extremely small compared to the amount needed to totally free the same mendicant electron from a bound state.

Movement of an electron from a regular bond to an impurity bond would leave a regular bond (one without impurity atoms) with a hole in it. This is like saying that the original hole in the impurity bond moved to the regular bond. Another nearby regular bond would soon repeat this movement—the hopping of a bound electron to a nearby hole—propagating the hole randomly through the lattice. Holes behave much the way free conduction band electrons behave. They move around freely for a period called their lifetime. Their lifetime ends when a free electron (of which there are few in a p-type material) falls into them. As we said, a semiconductor dominated by free holes is called p-type.

Since an impurity that adds extra free holes to a lattice accepts a nearby electron to create the free hole, such an impurity is called an acceptor. It actually becomes negatively charged. As above, n-type impurities are donors; p-type, acceptors. Together, they are called dopants, in the sense that their introduction in a relatively pure material alters its electronic properties.

Now we're ready to put these two kinds of semiconductors together. The result will be the essential element of a PV device: the fixed electric field that makes everything work.

Making a Built-in Field

It is important to remember that both n- and p-type semiconductors are electrically neutral. There are no more electrons than protons in either of them. Yet one can have a million-billion free electrons per cubic centimeter and very few free holes; the other a million-billion free holes and few free electrons. But charge neutrality remains unbroken in both conditions because the impurities in the n-type semiconductor have the same number of extra protons as the extra conduction electrons they introduce; and the p-type impurities lack just the right number of protons to match the number of extra holes they add.

Suppose we let n- and p-type semiconductors share a common boundary. Let's consider two neighboring rooms in our new building, one p-type and one n-type, both sharing a common wall. One room may have a bond missing an electron (the p-type room). But within its new neighbor there are many free electrons wandering around with no place to settle into a bond. Now, however, they see many possible sites at which they can settle down and lose some energy: they see holes in bonds in the p-type lattice right next door.

When n- and p-type semiconductors are juxtaposed, free electrons from the n-side wander across the boundary and settle comfortably in bonds on the p-side. This is how an electric field forms.

What is to stop this from going on until all the electrons from the n-side have crossed over and filled holes? Charge neutrality is being upset! As each electron crosses over, it leaves behind a positive charge; and as it settles into place, it adds a negative bias to the p-side. In a narrow region on both sides of the interface between the two materials, an electric field builds up. Each new electron that would pass from the crowded n-type side to fill an incomplete bond on the p-side finds it harder to climb the growing electric field. Finally, an equilibrium is reached, and net charge movement stops. The electric field is too high to climb for electrons passing from the n- to the p-side.

We have just described the formation of the heart and soul of PV, the cell's built-in electric field. To recapitulate, the proximity of holes to an n-type semiconductor teeming with free electrons allows some of the electrons to accidentally find empty holes on the p-type side. More would come, but each that does sets up a greater and greater charge imbalance—an electric field—that eventually prevents others from following them. Although the *bulk* of the n-type side (away from the electric field) still teems with electrons, and the p-type side still teems with holes, the electric field at their interface forms a barrier keeping them apart.

The electric field is not tenuous; it does not come and go. We call the area of the field the depletion region because in that region there are no free carriers—all are either in bonds or swept away by the field. The electric field in the depletion region is quite strong, capable of accelerating electric charges that pass through it to over a hundred kilometers per second.

Incidentally, what we have just described is simply a diode. If used in a circuit, it would let current flow easily in one direction and prevent it from flowing in the opposite direction. Electric current—electrons—would be blocked from flowing against the field (blocked from going from the n-type to the p-type region) but would be accelerated if they flowed the other way—with the field. Diodes are very familiar solid-state devices used in everything from clocks to stereos to battleships. PV devices are not exotic—they are simply light-driven diodes.

Let There Be Light

As we know, photons of sufficient energy absorbed in a semiconductor can free electrons from a bond in the lattice. We call this the formation of an electron–hole pair, because a free hole is created at the same time as a free electron.

Suppose this electron–hole pair were formed on the p-type side of an n–p electric field region. Remember, the p-type side has many free holes in it already, but almost no free electrons. The new, light-generated electron would have only a short time to move around randomly before it would fall back into one of the numerous holes on the p-type side. However, in that short time, the electron *might* encounter the built-in electric field. What would happen? The electric field opposes electrons flowing from the n-type region to the p-type region. On the other hand, it *favors* electrons moving from the p-type region to the n-type one. Our free electron is in the p-type region. If it encounters the electric field, it will be driven downhill across the interface to the other side at a hundred kilometers per second.

An analogous process occurs on the other side of the cell. A light-generated free hole formed (along with a free electron) on the n-type side might be pushed by the built-in field into the other, p-type lattice. How? Holes are the absence of electrons from bonds. Electrons in the lattice could hop from hole to hole across the boundary from the p-type to the n-type side to fill a newly created hole. This would be favored by the electric field just as the parallel movement of free electrons in the conduction band is favored.

The built-in electric field works in a symmetrical manner to propel electrons from the p-type side into the n-type side; and to propel holes from the n-type side into the p-type side. In effect, it separates the light-generated free charges (holes and electrons) formed within either the n- or p-type region of the cell.

Majority and Minority Carriers

We have called one region n-type because it is dominated by negative charges and the other p-type because it is dominated by positive holes. These charge carriers are called the majority carriers in their respective materials because they far outnumber the other kind of carrier within that material.

The built-in electric field is formed when some of the majority carriers move across the boundary and settle into fixed positions. Subsequently, movement of free charges (current) occurs when light generates electrons and holes. Then the *minority* carriers (electrons on the p-side; holes on the n-side) move oppositely: electrons from the p-type to the n-type side; holes from the n-type to the p-type side. During operation (in sunlight), *minority carriers carry the current*, moving from the side in which they are fewest to the side in which they predominate, i.e., electrons flow from the p-side to the n-side; holes flow from the n-side to the p-side. Both contribute to current flow.

Fortunately, it is not necessary to remember all these movements of electrons and holes to and from n- and p-type sides. (One handy memory device is that carriers always flow from the side in which they are in the minority to the one in which they are the majority.) The point is that our PV cell is set up to move free carriers (electric current) when it absorbs light. Whenever light is absorbed—as long as it's absorbed near enough to the electric field—free charges can be pushed across the interface by the electric field. A charge imbalance is created. By attaching a wire to the cell, we can have a continuous current of free charges seeking to reestablish charge equilibrium as long as light shines on the cell.

Theoretical Efficiencies

We have most of the pieces needed to estimate the theoretical efficiencies of different PV devices. In practice, the power output of a PV device is a multiple of

☐ The number of electrons and holes that get separated by the electric field (i.e., the electric current); and
☐ The amount of difficulty these separated charges would have climbing back up that electric field the wrong way (the voltage).

We can think of the flow of light-generated electrons (and holes) the way we think of water being pumped through a pipe. The amount of water is analogous to the number of electrons (the current), which is proportional to the amount of sunlight. The strength of the pump driving the water is analogous to the strength (voltage) of the electric field driving the electrons. The product of these two terms (current times voltage) is the power.

We would like to know how efficient solar cells are likely to be in changing light into electricity. To do so, we must be able to estimate the number of electrons that are participating in the current and the force (voltage) with which they are being propelled through the circuit.

Light and Current

We want to estimate how many photons contribute to the electric current. The concept of semiconductor *band gap* is essential for this estimate. Each semiconductor material (e.g., silicon, selenium, cadmium telluride) has a unique band gap. A material's band gap is simply the energy at which a photon will dislodge an outer electron from its bond. This characteristic energy varies for different semiconductor materials, because each of them has a different bond strength. Those whose bonds are weak have small energy band gaps (weak photons can dislodge electrons); those with strong bonds have high band gaps.

Semiconductors are almost totally transparent to photons that have less energy than their band gaps because this kind of light does not have enough energy to be absorbed. Such photons pass right through the semiconductor as if it were glass. But almost all light with more energy than needed to free an electron in a given semiconductor is absorbed by that semiconductor and generates free electrons and holes. Semiconductors can be very powerful absorbers of light. In some cases, only a micron (a ten-thousandth of a centimeter) is sufficient to absorb 99% of the above-band gap photons in sunlight.

To know how much current (flow of electrons) light generates in a cell, we must first know how many photons are producing electrons and holes in a given semiconductor.

Recall that sunlight is a spectrum of energies. Semiconductors are transparent to light that does not have enough energy to generate an electron–hole pair. If we wanted to absorb all of the solar photons, our semiconductor band gap would have to be very small. Then almost all of the sunlight would have sufficient energy to generate electrons and holes and would be absorbed. Wouldn't this be the right way to design a PV cell? Our currents would be as large as possible. Unfortunately, it doesn't work. We would get a lot of current; but the voltage—the oomph with which the current is driven—would be negligible. This is because—as we will soon find out—the electric field that drives our current is proportional to our material's band gap. A small band gap means a tiny voltage.

So why not have a giant band gap, producing a giant voltage? Because as we increase the band gap, more and more of the spectrum of solar photons lack the energy to produce electrons and holes. The semiconductor becomes transparent to a larger proportion of our sunlight. At the extreme, we would have a powerful pump (high voltage) pushing a dribble of water (miniscule current).

In practice, we must compromise and choose a band gap that allows us both reasonable current and reasonable oomph.

Voltage

We want to see how voltage relates to band gap. But first, let's see how voltage relates to the number of free carriers, because this is an even more basic fact of life in PV.

Voltage is directly related to the strength of the electric field between two semiconductors because this defines how high the hill is between the semiconductors. We have already seen how this field forms. The greater the density of free carriers initially on each side of the interface, the greater will be the electric field that forms when they mix. This is because initially more free electrons are around to fall into more holes. A greater electric field has a chance to form just because there are more nearby spots into which more electrons can accidentally fall.

Think of this like filling holes in a Chinese checkerboard with marbles. Suppose we have a checkerboard with a lot of holes. Obviously, it can hold a lot of marbles. More holes would be filled by our randomly moving marbles (electrons) before equilibrium set in. The electric field that forms is larger because it is proportional to the number of transferred electric charges.

This is a pretty basic effect. If two slightly n- and p-type semiconductors are in contact (i.e., ones with few free charges), not much of an electric field will result because not many charges will cross over. A weak electric field won't make a very good solar cell. If the materials are strongly n- and p-type (many free charges), however, a larger field forms and we have the chance of having a good PV cell.

This explains why greater numbers of free charges result in higher voltages. But why does the electric field (and voltage) increase with band gap?

Band gaps describe the amount of excess energy that is needed to free an electron from a bond. Larger band gaps mean stronger bonds. To continue with our Chinese checkers analogy, this is like saying that larger band gaps imply stickier (or deeper) holes in the checkerboard. For the same number of marbles, more of them stick in the holes. We get more marbles per square centimeter, and we get marbles that find holes further away from the source of the marbles just because each that does has a higher probability of sticking there. In terms of electrons, we get more electrons filling available holes and travelling further into the p-type material before the opposing field they create stops them. The voltage of PV cells is proportional to the electric field and the distance that field covers. Increasing the number of free carriers and increasing the band gaps of the n- and p-type semiconductors enhance the voltage by increasing the number of free carriers that cross the interface during the formation of the built-in field.

Efficiency Limits

Efficiency potentials can be projected for solar cells made of different materials based on the band gaps of the semiconductors in the cells. Different materials—different band gaps—have characteristically varied best possible theoretical efficiencies. However, these projections are only in terms of a simple, generic cell design. In fact, some rather complex, but doable, designs already exist for surpassing these simplified best efficiencies. For instance, Boeing Aerospace recently announced a 34%-efficient cell, which is well beyond the potential of simple PV devices. It would be hard to say how such an efficiency could be reached with the kind of design we are about to describe. However, simple devices may yet turn out to be the most relevant because they will cost less and be made more easily. So projections based on very simple designs are still valuable.

Others have done careful analyses of loss mechanisms controlling solar cells and have derived theoretical efficiencies for the simplest PV design. Figure 11 shows their results. Two things are evident: simple PV cells can be as much as 30% efficient, and a plateau of efficiencies above 20% exists between band gaps of 0.9 and 1.8 eV. There are many semiconductors with these band gaps, which is an important technological fact and part of the reason that there is a great richness in PV research.

Some have cast aspersions on PV as being inefficient and therefore unlikely to be economical. They consider typical PV efficiencies—10 to 25%—to be low compared to conventional power plant efficiencies. This is a completely erroneous viewpoint. They compare

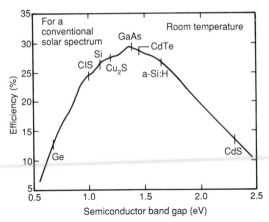

Figure 11. The theoretical efficiencies of absorber materials with band gap energies ranging from 0.5 to 2.5 eV. The peak of about 30% efficiency occurs at 1.4 eV, but there is a broad plateau of over 20% efficiency in which there are many suitable PV materials.

PV with conventional, fuel-using energy production—burning coal or oil, or nuclear energy—all of which can be 40% efficient. This is a most absurd and tendentious comparison. The relevant comparison would be with other forms of direct energy production. What was the original conversion efficiency of the process that produced fossil fuels? It was photosynthesis and then biological decay, certainly at less than 1% efficiency when compared to the energy in the original sunlight. In fact, *PV is perhaps the most efficient means of transforming the primeval fuel—sunlight—into electricity.* This is by no means a trivial advantage. It ultimately bears on the fact that in competition with other solar technologies—biomass growth and conversion, for instance—PV looks very efficient and attractive for large applications. For instance, land use for PV is about 20 times smaller than for biomass-based energy production. For this reason, PV will have a significant advantage in the long-term competition for the supply of a major proportion of global energy. Meanwhile, for the near term, the relevant comparison is the *economics* of the various methods of making electricity. We have already established the parameters of such a comparison in the previous chapter, and we have similarly established the goals that PV needs to achieve to compete with the cost of conventional alternatives.

Loss Mechanisms

Actually, we already have most of the tools needed to depict the loss mechanisms that limit the simplest PV devices to about 20–30% efficiency.. The major reason for all of these losses is the same: inefficient use of the solar spectrum by PV cells that have only one built-in electric field. Later we will see that complex designs with several built-in fields are possible. The 34% cell mentioned above had two built-in fields.

Inefficient use of the spectrum results from two things: (1) some proportion of the sunlight is not used because the photons do not have enough energy to be absorbed and to free electrons. (2) For those photons that do have enough energy, there is no discrimination between them—they are all treated as if they had just enough energy to free an electron and no more.

Let's see why. We already know that the higher a PV material's band gap, the more photons pass through it unused. Figure 12 shows the fraction of sunlight that materials of various band gaps are transparent to (ignoring sunlight below 1.0 eV). Losses from unused photons vary for different band gaps, but on the average are about 30% of the original energy in sunlight. Based on this alone, a simple PV cell's ultimate efficiency could not be more than about 70%.

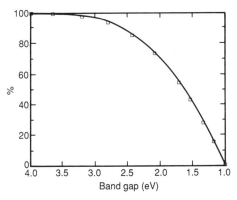

Figure 12. A semiconductor is transparent to photons with energies below the semiconductor's band gap. Semiconductors with high band gaps are transparent to most sunlight; those like silicon and copper indium diselenide with low band gaps absorb almost all the solar photons. If all the photons in the range 1.0–4 eV were transformed to electricity, they would provide about 500 amps of current from each square meter of surface area.

Now let's see about losses from light with *too much* (rather than too little) energy. What happens to light entering a low-band-gap material, say one with a band gap near 1.0 eV? Numerous photons are absorbed, and most have more energy than 1 eV. But no matter what their energy, the solar photons produce one electron–hole pair in the cell. In theory, light with more than *twice* the band gap energy might produce two electron–hole pairs (each taking 1 eV of energy), but in practice this is a negligible effect. The distribution of sunlight in the solar spectrum does not favor it because even for the smallest usable band gaps near 1.0 eV, there are very few solar photons with more than 2.0 eV energy. Because every photon with more energy than the band gap produces the same effect—one electron–hole pair—the PV cell has no way to use higher energy photons more efficiently than it uses· the ones at 1.0 eV. The low-energy photons produce one electron–hole pair, and so do the others. The cell has turned our sun's energy-rich spectrum into one where all the photons carry only 1.0 eV each, at least as far as the PV cell's output is concerned. The upshot is that low-band gap materials absorb almost all of the photons but use them poorly. The cell's band gap becomes the least common denominator of the cell in terms of its potential for using energy-rich photons.

Between these two effects (not using photons because they don't have enough energy; and using all of those that are absorbed as if they had one *low* energy equal to the band gap) we can account for a loss of about 70% of the solar energy. We cannot have devices over about 30% efficiency if they are configured with only one electric field.

Multijunctions

We now know enough to understand why we could expect much higher efficiencies if sunlight were monochromatic. We would simply design a cell with a band gap slightly below the energy of the mono-chromatic light and transform almost all of that light into electricity. The major losses—light that has too little energy and light with too much—would be gone. Our theoretical efficiency might be well above 80%.

But there is another—more practical—way to utilize the solar spectrum more efficiently. It requires a more advanced design than the cells we have described so far. Called a multijunction cell, such a design is based on using more than one electric field to separate electrons and holes. Figure 13 shows a schematic. A multijunction is basically one cell stacked on top of another. Light is incident on the top cell, which has a high band gap. The portion of the spectrum with energy more than the top cell band gap is absorbed in the top cell. If the top cell is properly made (i.e., if it has a transparent back contact), light with less energy than its band gap will go through it and be absorbed in the bottom cell. In this manner, the photons of the solar spectrum are split between two cells, the high-energy ones in the top cell and the low-energy ones in the bottom cell. In theory, we could have as many cells as we want in our stack; in practice, three is currently the practical limit.

We could think of this as merely putting another cell under one we already have. Clearly, if the top cell is transparent and the bottom cell has a lower energy band gap, we can gain some added power from such a combination. In fact, we can do better with two electric fields than we could ever do with just one. The reason is that we are splitting the solar spectrum so that its photons are used more efficiently. High-energy photons are used in a high-band-gap cell with a high voltage; low-energy photons go to a low-voltage cell. Losses due to the mismatch between the energies of the photons and the cell's band gap are reduced. Multijunctions come in two generic forms called two-terminal and four-terminal devices. In the two-terminal device, the stacked

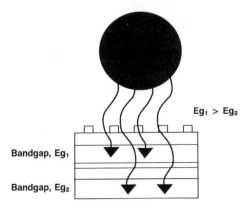

Figure 13. A two-junction cell can be thought of as the combination of two single-junction cells, with the light passing through the top one being used in the bottom one. With proper materials, two junction cells can be about 40% more efficient than either cell by itself.

cells are connected at a common boundary—the bottom contact of one is the top contact of the other. Current flows continuously between the cells under illumination. For optimal performance, the light-generated current of both cells must be very close. Otherwise, the lowest current will limit the entire device. Thus, band gaps must be chosen that split the spectrum equally, half of the sunlight absorbed on top, half transmitted to the bottom cell and absorbed there.

Four-terminal cells do not have a common boundary. Their output is taken off separately; each has a top and bottom contact connected to an external circuit. Their performance is independent, i.e., they do not have to split the spectrum between them to avoid performance penalties. In that sense, they are more easily optimized. In practice, other losses mount up, such as unwanted reflections between cells; and costs can be significantly higher for four-terminal cells. Usually some property of the materials—perhaps their sensitivity to process temperatures—determines which approach (two- or four-terminal) can be done. In comparison to single-junction cells, all multijunctions are more difficult to make and more expensive, which is why they are not universally adopted despite their higher efficiencies. Yet they remain a significant pathway for future PV development.

The theoretical maximum efficiency of the best two-junction cell is about 50% more than the best single-junction cell—i.e., about 40% efficiency. By stacking many cells—impractical with current technology, but theoretically possible—efficiencies as high as 70% or more would be possible. Perhaps deep in the next century, when the technical framework exists (e.g., when semiconductor processing technologies have advanced much further), these kind of efficiencies will be the norm. However, the cost-competitive position of PV does not depend on such an eventuality.

Conclusions

The band gap at which spectrum-driven losses are smallest in single-junction cells is about 1.4 eV (see Figure 11). However, a plateau of band gap values at which losses are still manageable extends from about 1.0 to 1.8 eV; many PV materials exist in this range and can be investigated for optimal PV cells. Between 1.0 and 1.8 eV, the theoretical sunlight-to-electricity efficiency limit of these simplest of all PV designs is 23–30%.

In terms of the near-term development of cost-effective PV, we can make the following observation. PV costs are based on the output (efficiency) of PV devices and the cost of their manufacture and assembly. With blood, sweat, and tears (and multijunctions), we may be able to double a device's efficiency—say from 10% to 20%. But if we need an order-of-magnitude reduction in electricity cost to make PV competitive by the turn of the century, the most fruitful avenue is likely to be reduced manufacturing costs. Conceivably we can lower those by a greater magnitude (more than five to one) than we can improve efficiencies.

On the other hand, from a very long-term perspective, if we want to reduce PV costs still further, we will eventually have to reemphasize efficiency improvements via exotic designs and vastly more sophisticated processing methods.

But for the moment, let's return to the behavior of existing PV devices. In the next chapters, we will continue to explore the design parameters of optimal solar cells. After that—when we understand where we can cut corners—we'll find ways to lessen manufacturing costs by significant amounts.

5 ☼ Device Parameters

Light enters a PV cell and generates electrons and holes, which are separated by an electric field. That's the simple story of a PV cell. But from the standpoint of practical devices, several key qualities of materials and design control the effectiveness of changing light into electricity.

Light Absorption

The ability of a semiconductor to absorb sunlight is a *critical* PV property. Some semiconductors are very efficient light absorbers. They can absorb nearly all (90%) sunlight in about a micron of thickness (10^{-4} cm). On the other hand, some semiconductors—even very useful ones—are much less capable of absorbing sunlight and may need as much as 100 microns (10^{-2} cm) to absorb the same fraction (90%) of sunlight. The difference in device design and material requirements for these two types of semiconductors—strong and weak light absorbers—is immense. Examples of very absorbent PV semiconductors are cadmium telluride, copper indium diselenide, and amorphous silicon. An important example of a poorly absorbent semiconductor is crystalline silicon.

Very strong light absorbers are called *direct* band gap semiconductors; poor absorbers are called *indirect* band gap materials. The difference in absorption strength between direct and indirect band gap semiconductors comes from the different processes by which they absorb individual photons. We described the general principles of light absorption earlier: When a photon enters a semiconductor, it can free an outer electron from a bond between neighboring atoms. We know that to do this, the photon has to have a certain requisite energy equal to the material's band gap. Light with less energy cannot interact with the bound electrons and passes through the semiconductor as if the semiconductor were transparent. We may think of semiconductors as a little like glass. They are transparent to below-band-gap light but absorb (and are opaque to) higher energy light. Glass has very similar properties. It looks transparent to us because the light it absorbs is invisible, ultraviolet light.

Different processes dominate the absorption of light in direct and indirect band gap semiconductors. In direct gap (highly absorbing) semiconductors, the process for light to generate electrons and holes is quite simple and is just like the generic process described above. Light of sufficient energy to free an electron from its fixed state is absorbed by the electron, which is then freed. We call these free electrons *conduction* band electrons to indicate that they can carry electric current (i.e., *conduct*). The light-absorption process is less straightforward in an indirect band gap (poorly absorbing) material. In this case, promotion of an electron to the conduction band requires the simultaneous interaction of a photon and a thermal vibration of the crystal lattice. The crystal vibrations are the natural result of heat within the lattice. They are called phonons. When a photon and phonon of the right energies are both absorbed simultaneously by the same bound electron, a free electron–hole pair results. This is the dominant process that takes place in indirect band gap semiconductors.

Indirect Band Gap Light Absorption

The physical phenomena behind the absorption of light in indirect band gap material are complex. For instance, almost all of the energy needed to generate the electron–hole pair is carried by the photon; the phonon's energy contribution is negligible. So what role does the phonon play? In the simplest sense, the phonon (lattice vibration) acts to catalyze the process. As an analogy, let us conjure up the picture of a wagon train crossing the American West in the 19th century. The phonon is like a helpful scout guiding the wagon train to its destination. Say the settlers on the wagon train have just enough water to travel one week. The scout (i.e., the phonon) guides them to the only trail that can bring them to water within ten days. Without their guide, the settlers would either get lost or take a less efficient path. Most likely they would run out of water and not get to their destination. This is the role of the phonon: to guide the reaction of light with a bound electron in such a way that the electron can be released from its fixed state. In physical terms, phonons have the property of being able to change the momentum of bound electrons, allowing them to be freed from their bonds by the absorption of a photon.

We might think of the band gap of an indirect semiconductor as actually quite a bit larger than its stated value. The conventional value assigned to an indirect semiconductor assumes the phonon interaction. For light to free bound electrons in an indirect band gap material *without* the help of phonons (direction changers), substantially more energized photons would be required. For instance, crystalline silicon's indirect band gap is 1.1 eV. But it takes photons of about 3.5 eV to produce an electron–hole pair in crystalline silicon *without* the aid of a phonon. Without its indirect band gap transition, crystalline silicon would absorb almost no sunlight and would be almost useless as a PV material.

Absorption Length

The freeing of a bound electron in an indirect band gap material requires two interactions—a photon and a phonon. But the simultaneous presence of a photon and phonon of the right energies is a much lower probability event than the presence of the photon alone. Above-band-gap light has a much greater chance of penetrating deeply into an indirect gap material before it is absorbed.

The average distance that light travels into a material before about 63% of it is absorbed is called the absorption length of that material. The absorption length characterizes how strongly a material absorbs light. It also is a critical parameter used in designing cells because it partially defines how thick to make the absorbing layers. The absorption length of light in a semiconductor is determined by the probability that a photon will be absorbed. For a particular material, a photon has the same probability of being absorbed in the first layer of a given thickness as in the next one. Thus, the probability of the photon penetrating two layers is the square of the probability of it penetrating one layer. So its chances of penetrating many absorption lengths is worse than the probability of flipping many heads in a row—it quickly becomes quite low. For instance, after four absorption lengths, almost all (98%) of the light has been absorbed. The absorption length for crystalline silicon, a poorly light-absorbing, indirect gap material, is about 30 microns. For direct band gap semiconductors like cadmium telluride and copper indium diselenide, the absorption length is much smaller: about 0.3 micron. Above-band-gap light does not penetrate very far into these materials.

Dependence on Photon Energy

Actually, the concept of absorption length is complicated by the fact that photons of different energy have different absorption lengths even within a given material. Higher energy photons have a greater probability of interacting with a bound electron and being absorbed than lower energy photons, even though both have more than the band gap energy. For instance, high-energy light (ultraviolet light) may have ten times the chance of being absorbed as infrared light (assuming both have more energy than the band gap). This means that in some materials the low-energy light will penetrate about ten times further before it is absorbed. For instance, in copper indium diselenide (band gap 1.0 eV), the absorption length of high-energy photons (energies above 2.5 eV) is less than 0.1 micron; but a beam of low-energy photons (1.1 eV) might require 1 micron to be equally attenuated. Even for crystalline silicon (band gap 1.1 eV), the difference can be quite large. Photons at 2.5 eV are absorbed in 0.5 micron (1 absorption length); but at 1.2 eV, photons take 100 microns to be equally absorbed. The differences between the absorption of high- and low-energy photons have an impact on cell design and cell performance limitations.

Placing the Electric Field

A semiconductor's absorption length is the critical parameter controlling PV device design. We know that the key action in a PV device is the separation of electrons and holes by a built-in field. For this separation to occur, we know that the light-generated minority carriers have to be near enough to the built-in field for the field's influence to send them to the other side of the PV device. We have not said very much about how close they have to be for this to happen. In fact, proper placement of the electric field in relation to the absorbed light is the most crucial aspect of an effective PV device.

No matter what we do, light of different energies will be absorbed at different depths in a semiconductor. This means that we cannot place the field perfectly—at least in the sense that some light absorption will always be more distant than we desire. The best we can do is to place the electric field in the middle of the area where most sunlight absorption is taking place. If we do so, much of the sunlight will be absorbed and will generate free carriers in the right place, near the field. Still—as we have just found—some of the high-energy sunlight will be absorbed in front of the field; and some of the low-energy sunlight behind it will go deeply into the material before being absorbed. This is an inevitable mismatch between the location of the field and some of our sunlight-generated electrons and holes. We know we can take advantage of free carriers that come within the influence of the electric field. But can we make use of the distant, light-generated free carriers to produce electricity? The answer is that we try, by using two phenomena: drift and diffusion.

Drift

As in the previous example, let us suppose that the built-in field is located at the center of the distribution of absorbed sunlight. Much of the design and fabrication efforts of PV are aimed at achieving this placement of the electric field. The built-in field has an important role in the separation of free carriers. The dimensions of the built-in electric field region are quite small (about a micron in width) *but not negligible*. If we do our designing right, some—even most—of our sunlight can be absorbed *within the field region itself*.

To the extent that this is possible, it is extremely valuable for achieving very high electric currents. Almost every electron–hole pair generated by light absorbed within the electric field region could be separated and contribute to the current. The light-generated electrons would be sent one way by the field, and their companion holes would be accelerated the other way. This would provide near-perfect separation of electrons and holes. As stated previously, an electron accelerated by the built-in field will attain velocities greater than 100 km/s. The field-driven movement of free carriers within the built-in field is called drift. The built-in field is a very powerful separator of oppositely charged free carriers once they are within its influence. In direct band gap (strongly absorbing) materials, most of the photons in sunlight may be absorbed within, or very close to the electric field, because typical absorption lengths (1 micron) in these materials are the same as the width of the built-in field. With the right placement of the field, nearly all of the sunlight-generated carriers are automatically separated and contribute to the electric current. The ease with which a properly placed electric field in a direct gap semiconductor can lead to near-perfect electron–hole separation is a big advantage for solar cells based on these materials.

But even in a direct band gap material, some light may be absorbed outside of the field region. And in *indirect* band gap materials—with absorption lengths of 10 microns or more—almost all of it is. Another mechanism—called diffusion—separates these charges.

Diffusion

Fortunately, electrons and holes freed by the absorption of light do not instantly lose their energy and fall back into bound states. The amount of time they remain mobile is called their lifetime. Consider an electron generated by light outside of the built-in field region in a p-type (hole-dominated) material. Unfortunately, the electron has lots of opportunities to fall back into one of the numerous holes that are teeming on the p-type side. If it does lose its energy (as heat) and returns to a fixed state in an empty bond, it is said to have recombined. But for some small time, the lifetime, the free electrons on the p-side move around randomly before recombining. During this short time (usually milliseconds to microseconds), they have a finite chance of encountering the built-in field and being accelerated to the other, n-type side of the device. If they do get separated, it is said that they did so by diffusion—a random motion that brings them accidentally into the vicinity of the electric field.

The average distance that light-generated minority carriers (electrons on the p-type side; holes on the n-side) can move toward the built-in field before they drop back into fixed states is called the diffusion length. The longer the diffusion length of a material, the greater is the chance that it will have a superior current. Diffusion lengths in real PV materials vary from less than a micron in most thin films to over 100 microns in some single-crystal materials. In a material with a 100-micron diffusion length, light absorbed at 100 microns from the built-in field still has a good chance (about 63%) of contributing free carriers to the electric current.

Good diffusion lengths are absolutely necessary for the successful performance of indirect band gap materials like crystalline silicon. Almost all of the sunlight absorbed within indirect gap semiconductors produces free charges far from the electric field. Without a good diffusion length, materials like silicon would produce very little current because most of the free carriers would recombine. On the other hand, *direct band gap* materials absorb so much light within their electric fields that their performance is much less affected by short diffusion lengths. In truth, the key parameter controlling diffusion-driven current collection is a material's diffusion length as a fraction of its absorption length. Materials that absorb most sunlight within a very thin layer can tolerate relatively small diffusion lengths. Thus, if the ratio of diffusion length to absorption length is greater than one, most carriers will be separated. For most direct band gap materials, whose absorption length is under 1 micron, diffusion length is of secondary importance since most photons will be absorbed within the electric field.

Diffusion lengths are weakly dependent on the number of majority carriers in the sense that the more majority carriers there are, the more opportunities will exist for the light-generated minority carriers to find one to recombine with. But far more important in determining a material's diffusion length is the quality of its crystal lattice.

Recombination Centers

Diffusion is limited by the tendency of free carriers to recombine. The simplest version of recombination is when a minority carrier—

say an electron in a p-type material — accidentally encounters and falls into a hole. Actually, this is a relatively rare event even when there are many holes available. The materials where this recombination mechanism dominates are actually superior materials that have extremely long diffusion lengths.

A much more frequent scenario is when a free minority carrier encounters something called a recombination center. Recombination centers *mediate* recombination. A crystal lattice has impurities in it — those which were intentionally added to make it n- or p-type and those which were inadvertently present due to impurities in feedstock materials. The latter exist because it is almost impossible to completely refine starting materials enough to eliminate all impurities, and semiconductors can be sensitive to impurities in the parts per billion range.

Although purposely added impurities may be essential in making a material n- or p-type, they do not sit perfectly in the lattice the way the regular atoms do. Their presence strains and warps the lattice. Lattice strains caused by these donor or acceptor atoms act as recombination centers in the sense that the random movement of a free electron (or hole) might be interrupted by an encounter with them, causing the free carrier to lose energy. Having lost some energy, a free carrier becomes much more vulnerable to reverting to a bound state.

However, these kinds of recombination centers are not the most serious. Other, unintentional impurities (such as iron in silicon, or carbon in cadmium telluride) are much worse. Located in various awkward positions in the lattice, they cause large strains that are very likely to impede the flow of free carriers. But the worst effects occur because most impurity atoms do not even find sites in the lattice; they sit in weakly bound positions between lattice sites. They can even hop from site to site and generally get in the way. These kind of impurities are called *interstitials*. They are very effective recombination centers.

Unfortunately, even highly pure materials can have interstitials. For instance, when dopant impurity atoms are purposely added to control a material's electronic properties, some or most of them may not find sites in the lattice. Instead, they may lodge between other atoms that are already filling the lattice's proper sites. To some extent, this occurs during doping in all semiconductors. But in those that are particularly hard to dope, up to 98% of the impurity atoms may inadvertently become interstitials.

Another example of an inadvertent problem in relatively pure materials occurs when one element of a compound (say Te in CdTe) is present in excess of the other. Pure compounds, like perfect lattices, usually have superior electronic properties. The imbalance between the elements in such a compound can be very small, say parts per million, and still cause serious consequences. One is that the element may precipitate from the lattice as metallic impurities, causing microscopic shorts. Another is that the imbalance may control the electronic properties of the material, making it impossible to alter favorably. Such semiconductors are said to be *compensated* because their electronic properties cannot be changed by introducing new impurity atoms.

Infrequently, however, the consequences are good of having too much or too little of an element. For instance, some two-component compounds, including CdTe, can be made p-type by adding a *very small* excess of tellurium atoms. The process must be carefully controlled to avoid the precipitation of tellurium metal (as described above). But tellurium atoms that substitute for cadmium atoms in the CdTe lattice lack two electrons to complete their outer shell. They attract other electrons. When they are in a lattice position, they grab excess electrons (perhaps from other impurities) and make the material less n-type and—by default—more p-type. However, such a Te-rich CdTe lacks some Cd atoms in their proper places in the lattice. These empty lattice sites are called vacancies and provide another route for increased recombination.

Other Defects

Numerous other possible defects can occur in the lattice and contribute to recombination, thereby reducing diffusion lengths. Several could be called growth defects. These are not impurity atoms but actual defects in the lattice that occur during its formation. They can be likened to a building with poor workmanship. Bricks may be absent or turned sideways; joists may not be firmly nailed together. In a semiconductor, structural defects like these can occur during the formation of the lattice. They can occur on a small scale—the absence of an atom in an expected place in the lattice (i.e., a vacancy)—or on a large scale—the complete fracturing of bonds along an entire surface.

This latter is called a grain boundary (Figure 14). Grain bound-

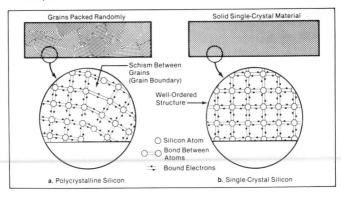

Figure 14. Except under near-perfect conditions, crystals such as silicon do not grow as single crystals. Usually, a multigrained (or polycrystalline) structure forms, wherein many small grains are packed together with numerous fracture areas, called grain boundaries. Grain boundaries are sites for many recombination centers that affect cell performance.

aries occur when a lattice takes shape in such a way that a defect perpetuates and makes it impossible for nearby atoms to bond together. The defect then expands greatly. Finally, the fault becomes so extensive that large, individual lattice structures, or crystals, grow next to each other but do not fuse. You can think of them as two lattice skyscrapers next to each other. If their walls were parallel and their corners lined up, they could merge into one skyscraper. But if their walls were to grow at an angle to each other, their corners could not line up: they would grow as separate, crystalline grains.

It is very common for crystal defects to cause this kind of multi-grained, or polycrystalline structure. Growing crystals that are all one grain—i.e., growing a single crystal—can actually be quite difficult and costly. Unfortunately, the borders between grains, the grain boundaries, can be the location of huge numbers of recombination centers. The bonds in these locations are all distorted from their normal character within a perfect lattice. In terms of their influence on cell performance, a free, minority carrier electron or hole that has to cross a grain boundary to get to the built-in field has less chance of doing so. Cell performance would suffer noticeably because these lost electrons would not contribute to the cell current.

There are some strategies for minimizing this problem in cells where we cannot avoid—say for cost reasons—the presence of grain boundaries. One approach is to *passivate* the grain boundaries. This consists of adding some extra material—hydrogen or oxygen are the usual candidates—that can make the grain boundary defects less harmful. Most added materials diffuse preferentially down grain boundaries, since those are easier paths for atomic movement than is the solid lattice. Thus, the added material goes directly to the region (the grain boundary) where it can have the most effect. For instance, putting a semiconductor in a furnace at 400° F can cause oxygen to enter it along its grain boundaries. The added oxygen might passivate defects in the grain boundary by grabbing loosely bound electrons, removing many of them as undesirable recombination centers.

Another approach to minimizing grain boundary effects is to grow the material in such a way that tall, columnar grains grow from the bottom of a cell to its top. Such columnar grains would look like Greek columns packed together into a solid. Grown properly, they would be perpendicular to the built-in field. Thus, a diffusing minority carrier would have the opportunity of reaching the electric field without encountering a grain boundary. A small fraction would still move randomly sideways and encounter a boundary (the surface of a column), but the majority would move vertically and encounter the electric field.

Finally, some materials have relatively harmless grain boundaries. The bonds within the boundary might be self-passivating; or some fortuitous step during semiconductor growth (say the inadvertent presence of oxygen) might passivate the grain surfaces. Materials with few grain boundary problems are particularly usable in cells where grain boundaries cannot be avoided. This is an important class of potentially lower cost materials. Over the years, various materials have been identified with this property of having relatively harmless grain boundaries. Two of them—copper indium diselenide and cadmium telluride—have become very important because they can be made inexpensively yet behave almost as if they were single crystals. Others with very noxious grain boundaries have also been identified, with gallium arsenide and silicon being among them. These materials require columnar grains and rather extensive passivation to make them at all usable in non-single-crystal form.

In general, inadvertent recombination centers (impurity interstitials and grain boundaries) thoroughly dominate the lifetime and diffusion length of free carriers in a semiconductor. But they are also relatively amenable to control via various improved growth strategies.

Strategies are continually being adopted to minimize them, which is one of the reasons that PV efficiencies have improved so dramatically over the years.

Shunts

So far, most of what we have discussed affects electric current: the number of light-generated electrons and holes contributing to current flow via drift or diffusion. But the concept of recombination centers naturally leads us into a different area as well: voltage losses — and these can be even more serious than current losses due to recombination.

A PV cell's voltage is essentially dependent on the height and width of its built-in field. But recombination centers located within the field can severely reduce the field's strength. A field is like a dike holding back water. Freely moving, light-generated charges that have already been separated by the field would like to pass through a hole in the dike to return to their original side and reestablish charge neutrality. But if the dike is high and has no holes, the free charges have to behave as we want, flowing through an external circuit (doing work) to get back where they started.

Recombination centers in the field region (called shunts) are like holes in a dike. They allow leakage, which reduces the effectiveness of the dike. The force — or voltage — with which electrons are propelled through the external circuit is diminished.

How do recombination centers result in defects in the dike?

Recombination centers are — by definition — capable of mediating the recombination of electrons and holes. In the bulk of the semiconductor, away from the electric field, they may interact accidentally with passing free charges. Each time a minority carrier recombines via such a center, the current of the cell is reduced below its optimum by one free carrier.

This is bad enough. But when a recombination center is *in* the electric field region, it is *always* causing recombination. It is like a hole in a dike, with water constantly rushing through it. Unlike a hole elsewhere, one hole in the dike can cause huge numbers of electrons to be lost. And worse than the *current* loss resulting from this is the fact that the height of the dike is effectively lower: *voltage* is markedly reduced by such shunts.

Each of the physical defects described above (e.g., grain boundaries, interstitials, vacancies) can be present in the electric field region and can act as a recombination center there. Unfortunately, another whole set of defects that are usually found only in the electric field region can also be present. They are special defects at the boundary, or interface, between the n- and p-type semiconductors, i.e., right in the middle of the electric field.

These kind of defects are not so much of a problem when the n- and p-type regions are in one material (such as silicon) doped with two different impurities. Then the only unusual increase in defects is because of the presence of two impurities rather than one. But when the n- and p-type regions are made of completely different materials (as they are in many cells), the problem can get a bit more serious. In such cases, an interface exists between the two materials—right in the center of the built-in field—where the lattice of one ends and the other starts. If one lattice is grown on top of the other during cell fabrication, there are likely to be structural faults starting right at the interface. Usually the shape of one lattice is a bit different from the other, and structural defects are unavoidable when one grows on top of the other. Then a multiplicity of recombination centers are introduced just where they can be most harmful: the center of the built-in field. In the worst extreme, the presence of numerous interface shunts is enough to destroy the field's effectiveness completely. Like a dike with too many holes, the cell is useless.

Voltage and Temperature

In addition to defects, several other phenomena can cause voltage loss. One of them is higher temperatures during actual operation. Voltage is like the height of a dike, in this case an unusual dike capable of keeping electrons in the n-type region and holes in the p-type region. The electric field holds the majority carriers apart. But it turns out that the relative size of the dike—or the electric field—is smaller at higher temperatures. Voltage is less. Why?

At low temperatures, the average random energy of electrons on the n-side is small. They have some energy to move around—like gas molecules in a balloon—but not very much in relation to the size of the electric field (the dike). A few of the electrons assume enough random energy to enable them accidentally to penetrate the electric field, but these instances are rare. At low temperatures, the dike appears to be very high; e.g., the voltage is a maximum.

However, as temperature rises (say, during use outdoors, as the sunlight heats the cell), more electrons on the n-side gain enough thermal energy to penetrate the opposing electric field and go over to the p-side. The dike appears less formidable to the average electron; this means that voltage is lower. In fact, if the temperature rises enough (to about 250–500° F), almost all the electrons will have enough thermal energy to go through the field as if it were not there. The device would essentially lose its character as a solar cell. Light-generated electrons and holes would have the same thermal energy as the others and would behave the same way—randomly. The cell would have no voltage.

Series Resistance

Another, but far more mundane voltage loss mechanism is series resistance. This loss occurs because of the same resistive loss we associate with any movement of free charges through wires. Any flow of current (electrons) engenders some frictionlike losses called resistance. The product of current flow times the resistance is the amount of voltage lost through resistance. Higher currents and higher resistances cause higher voltage losses.

The flow of electrons in wires is very efficient because wires are made of metals like copper and aluminum that have very little resistance. But significant resistance losses can occur in PV devices in the internal, semiconductor layers of the cell. Semiconductors can be millions of times more resistive than metals. Losses occur because free carriers separated by the built-in field must still travel some distance to reach a metal contact. At the back of the cell (where a metal layer covers the whole exposed surface), this distance may be purely vertical and quite small—maybe less than a micron. But the front contact is usually a grid to avoid shadowing. To reach the grid fingers, carriers must move sideways long distances—several millimeters— and can lose substantial energy to resistance in the process.

This resistance can be reduced by making the top layer from a semiconductor with a high density of free carriers. These kinds of semiconductors can have very low resistances—in the extreme, almost approaching those of metals. We have mentioned this kind of transparent, conductive semiconductor before.

The reason that a high carrier concentration helps conduction is as follows. Current does not flow as we may think, by the meandering of a single electron from one side of a cell to another and then out through a contact, etc. Instead, each time a new carrier—let's say an electron—is added (by field separation) near the bottom of a semiconductor layer, its presence helps push another electron out of the top of the same semiconductor into a metal contact. Even in resistive materials, the added free carrier does not actually cross the whole semiconductor to contribute to the current. It just pushes one of the other free electrons out the other side. But in materials with huge numbers of free carriers there are more of them to push out of the material, and the average distance that each carrier travels is smaller. Since resistance is proportional to the distance the carriers travel, total resistance is smaller when there are more free carriers. This is the reason that making the top layer with a lot of free carriers minimizes resistance losses there.

All PV materials have another added resistance called contact resistance. It is caused by an inadvertent barrier between the semiconductor and the contact metal. Like the induced electric field between two semiconductors, semiconductors and metals also induce an electric field. In some cases, it opposes the flow of current from the cell to the metal. Such electric fields can be quite severe in some semiconductor/metal contacts. Instead of charges flowing freely from the semiconductor to the metal contact, they resist their passage. Each carrier needs an extra push to get across. This contact resistance becomes a voltage loss that reduces the cell's overall voltage somewhat. In certain semiconductors, e.g., cadmium telluride, it can be a critical problem.

We now know enough about the details of how cells work to understand something about how their performance is measured and analyzed. This entails several procedures:

Quantum Efficiency

A conceptually simple but extremely useful measurement of solar cell performance is called the quantum efficiency (QE). It is a measure of the effectiveness of a cell in converting light of various energies into electricity. We know that light of different energy is absorbed at different depths of a PV cell. The light-generated carriers naturally have different chances of being separated by the electric field, depending on where they start in relation to the field. The QE technique measures the fraction of the carriers that contribute to the electric current.

To do the QE measurement, the cell is illuminated with monochromatic light while the cell's electrical output is being recorded. The experiment is set up so that we know the number of photons in the monochromatic light. We measure the resulting electric current, which tells us (via simple calculations) how many electrons are being produced by the cell. From these two amounts (the number of incident photons and the number of electrons leaving the cell), we can easily calculate the probability with which the photons in our monochromatic beam contribute electrons to the current. Then, by slowly changing the monochromatic light to various energies, we can measure the cell's response to the spectrum of solar photons.

The QE measurement can tell us a great deal about a cell. For instance, if the photons making up the monochromatic light have less energy than the cell's band gap, they will pass through it without producing any current. The QE of photons of this energy will be zero. Just above the band gap of the cell, light will be very weakly absorbed. Such photons are likely to produce only a small number of electron–hole pairs, and those that are produced are far from the electric field. Unless the material's diffusion length is very great, the QE will be small.

But the QE will begin to rise steeply as the energy of the incident photons is increased. Light in the middle of the solar spectrum is likely to be absorbed near the electric field in a well-designed device. Most of this light will contribute to the current. In some cells, QE can reach 95% or more for this kind of light (Figure 15). In fact, in very good cells (such as the one in Figure 15), QE of over 90% can be reached across most of the solar spectrum. This is especially true for direct band gap cells, because most of the sunlight can be absorbed in or very near the electric field.

A QE measurement is a simple analytical procedure that is often used as a diagnostic tool to characterize the problems of experimental cells. For instance, QE measurement of a cell might show that it had a very fine response to high-energy photons but a poor low-energy response. If this is the case, it is usually because the cell is not using the light absorbed far from its electric field. The cell's diffusion length needs to be improved by improving the structure of the absorber material or perhaps through passivating defects.

On the other hand, a cell with a poor *high-energy* response usually has an obstructing layer that is either opaque at these energies (but transparent at lower energies) or reflects high-energy light preferentially. If this is discovered via a QE measurement, the opaque or reflecting layer can usually be altered or removed.

Cells with a mediocre response across the whole energy spectrum may have several problems. The most obvious is that they may be reflecting a large fraction of the incident light. Another is that there may be large recombination losses in the junction that are removing a fraction of the electrons independent of the energy of their original photons.

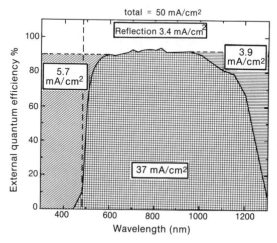

Figure 15. The quantum efficiency of a Boeing Aerospace Corporation CuInSe$_2$ cell. The quantum efficiency measurement shows what fraction of incoming photons contributes to a cell's electric current. It also shows where losses are occurring.

Figure 15 can be used to analyze the Boeing copper indium diselenide ($CuInSe_2$) cell on which the measurement was taken. The cross-hatched portion of Figure 15 is the portion of the spectrum contributing to the density of current (37 mA/cm^2). In this region of the spectrum (wavelengths of 500–1200 nm, or 0.5–1.2 microns) almost 90% of all photons contribute an electron to the current. The total possible current if all photons contributed one electron is about 50 mA/cm^2. Of that, about 3.4 mA/cm^2 is being reflected (top, unshaded portion above curve). Two other major current losses are 5.7 mA/cm^2 in the short-wavelength (high energy) photon region and 3.9 mA/cm^2 in the long-wavelength (low energy) region. The high-energy photons were lost because the cell had a thick cadmium sulfide (CdS) n-type top layer that absorbed most of the short-wavelength light too far from the electric field to be of use. The same holds for the long-wavelength losses: those are from low-energy light absorbed deep within the copper indium diselenide—so deep that most of the free electrons caused by that light do not get to the junction. Thus, we see that this cell uses about 74% of the light it could use; and we know where and why it loses almost all the rest. As we shall see in Chapter 10, the cell has subsequently been redesigned and its response to the spectrum substantially improved. Designs based on improvements resulting from analytic tools like QE can allow cells to be designed with near-perfect electric currents.

We already know that cells with lower band gaps can absorb more sunlight than those with higher band gaps. Table 5-1 shows the theoretical currents of cells with different band gaps if their QE were unity. Although this is impossible, good cells can have currents that are 90%

Table 5-1. Theoretical Maximum Currents for Different Materials[a]

Material	Band gap (eV)	Maximum current (mA/cm^2)
$CuInSe_2$	0.98	50
Silicon	1.1	43.4
GaAs	1.4	31.8
CdTe	1.5	28.5
Amorphous Si	1.7	21.7

[a]Assumes 100% quantum efficiency; band gaps and potential currents in actual materials can vary slightly from those given here.

or more of the theoretical value; and this mark provides a goal for designing optimal PV cells.

Current–Voltage Characteristics

PV cells provide energy via the flow of electric current through an external circuit. Electric currents are described in terms of their current and their voltage, the product of which is their power output. PV cells have current–voltage characteristics that are used to characterize them.

To understand a PV cell's current–voltage characteristics, we must begin with an extreme—no current at all. Suppose that a circuit connecting a PV cell to a load is open, i.e., no current can flow because the external circuit is disconnected. If light shines on the cell, carriers will flow across the electric field and build up on the other side of the cell. They will not leave the cell through a contact as they would normally. The buildup of charges will keep going until the added free carriers just balances the force of the cell's built-in electric field. Then the built-in field will be unable to separate any more free charges. This maximum buildup of separated charge is called the open-circuit voltage. Note that the voltage is actually proportional to the amount of charge backed up behind the built-in field. The voltage created during open-circuit conditions is the largest voltage the cell could generate, and it is measured by allowing an illuminated cell to produce no current.

Now we will complete the circuit but in such a way that only a small amount of current will flow. Suppose that the external circuit is completed by putting a very resistive load in it. Resistive loads require a high voltage to drive even a small amount of current through them. With an extremely resistive load in the way, the cell will only be able to push a very small amount of current through the external circuit. But this small flow of current will lower the amount of charge backed up against the electric field. The voltage we measure under such circumstances will be slightly lower than the cell's open-circuit voltage.

Let us continuously reduce the resistance of the load. As we do, more current will flow—but the backed-up charges will be fewer, and consequently the voltage will be lower.

At the other extreme, let us consider a PV cell providing power to a short circuit made of a nearly resistanceless wire. We will find that in this case we have plenty of current but not much voltage. Why? Each time a solar photon enters the cell and sends an electron over the built-in field, the electron easily flows through the nearly resistance-free external circuit. No charge builds up behind the dam, i.e., voltage is very low. But current is maximum: every separated charge provides a contribution to the electric current. This current is called the short-circuit current and is the maximum current that the cell can generate.

The *power* that a cell provides is the product of its operating current and voltage. Under short-circuit conditions, current is maximum but voltage is nearly zero: almost no power is provided to the circuit. At the other extreme, under open circuit conditions, voltage is highest but no current flows: power is again zero. Somewhere between these extremes is a condition called maximum power: it is a compromise in which both current and voltage are some relatively large fraction (about 70–90%) of their maximums. Cells, and the arrays made from them, are designed to produce electricity while they are near their points of maximum power.

The maximum power can never equal the product of the open-circuit voltage and the short-circuit current. But the closeness with which maximum power approaches this product is called the fill factor. Fill factor is a measure of how close a cell comes to behaving perfectly. Some cells can have good open-circuit voltage, good short-circuit current, but a very poor fill factor. The result is not much power and low efficiency. Mathematically, fill factor is defined as the ratio of the power at maximum power to the product of the short-circuit current and the open-circuit voltage. In good cells it is usually over 70%.

Efficiency Measurements

Efficiency is the measure of how well a cell works. In particular, it is the ratio of the power a cell produces to the amount of sunlight that is shining on it.

In practice, it is not very easy to make an accurate efficiency measurement. One reason is because sunlight changes under the influence of various conditions. The factors that influence sunlight the most are the angle in the sky from which the sunlight is shining, cloudiness, water vapor and other invisible components of the atmosphere, and the altitude of the site (since there is less intervening air to attenuate sunlight at higher altitudes).

In practice, the problem is one of defining a standard condition under which to measure all solar cells so that fair comparisons can be made. There is no such thing as typical sunlight except in outer space. Instead, to get a conventional measurement, we have to *define* a standard solar spectrum based on globally average conditions of sun angle, water vapor, atmospheric turbulence, etc., and then combine these influences into a fictitious but useful standard spectrum. Then we can use this standard spectrum as a basis for measurements.

The Solar Energy Research Institute has been responsible for introducing such a conventional spectrum. It is called the standard global air-mass 1.5 spectrum. It is a global spectrum in the sense that it includes all the sunlight—both the direct beam and the diffuse sunlight from the rest of the sky. Air mass 1.5 refers to the relative atmospheric thickness assumed—in this case, 1.5 times the thickness of the atmosphere (when it is measured at an angle normal to the Earth's surface). Air mass thickness is an equivalent way to define the sun's angle. At solar noon during the spring solstice, sun angle is 0° (straight up) and air mass is 1.0. At a certain sun angle (actually 48.2° from the normal), the equivalent of 1.5 of these noontime air masses (AM 1.5) is attenuating the sunlight. This angle (or air mass) has been chosen as more typical of operational circumstances than is an air mass 1.0 spectrum and is used in the standard measurement.

The fictional standard spectrum may or may not ever actually occur outdoors. However, it is very close to an average outdoor spectrum.

To measure the efficiency of a cell under this standard SERI spectrum, one must actually perform the measurement indoors under a solar simulator—equipment designed to mimic the SERI solar spectrum.

The heart of a simulator is a quality light bulb—for example, a xenon bulb—capable of repeated, controlled use. Its spectrum can be fairly close to the standard solar spectrum, but normally has some unwanted spikes and variations. Filters are then used to remove the spikes and bring the xenon spectrum closer to the standard.

To make a good measurement, two other parameters are controlled: total power in the light and the temperature of the cell. The xenon bulb can be regulated and focused so that it shines 100 milliwatts (mW) of light per square centimeter (equivalent to 1000 W/m^2) on the cell to be measured. This is the standard power to which all measurements are referred; and it is approximately the power density of sunlight at the Earth's surface at noon on a cloudless day. Meanwhile, the cell is held at a fixed, agreed upon temperature—77° F (25° C). The temperature must be fixed because cell voltage—and thus power output—varies with temperature.

The final necessary parameter for efficiency measurement is the cell area. To get efficiency, one must have a precise idea of cell area, because the amount of input sunlight depends on the cell area. For instance, if 100 mW/cm^2 is shown on 1 cm^2, then the amount of sunlight is 100 mW. This then becomes the denominator in the efficiency (with cell output in watts as the numerator). A mistake of 10% in the cell area leads to a similar mistake in the efficiency.

This kind of error does not sound like much, but in practice variations in area measurements have been the source of some of the most heated arguments in PV. Although various conventions have been adopted for defining cell area, a lack of precision in using these definitions has led to efficiencies that differ by 10–40%, or more! Usually the problem is defining where a cell ends. If a small, experimental cell is made on a larger substrate, where does the cell stop and the substrate begin? Precise measurements can be done, but sometimes they are not. Mistakes will inevitably follow.

It is essential to know what area definition is being used, especially when comparing different cells. Inconsistent assumptions can lead to very misleading results. Two prevalent definitions are based on concepts of total area and active area. Active area is the surface area of the cell without counting metal contacts and pads even if those are on top of the active portion of the cell. Total area is the total area of the top surface of the cell, including any contact pads that cover active parts of the cell. Even the contact areas that extend off the active area of the cell count. But inactive substrates that have no electrical connection with the cell do not count; nor do electrical contacts on the back of the cell. Active area is a smaller area than total area and leads to better efficiency numbers.

In the case of small, experimental cells, the active area definition may be more relevant because it helps to characterize the material and device design rather than the engineering of the grids and contact pads, both of which might be very differently shaped for larger area devices. For modules, total area (including frame) is appropriate because modules are commercial products; the consumer wants to know how much power is produced per unit area, no matter what that area consists of. However, another area convention applies to *submodules*—small, experimental modules. This convention is called aperture area efficiency, and it consists of measuring the area inside of the submodule frame. The rationale is that when these submodules are scaled up to larger, commercial products, the fractional area associated with the frame will be smaller; including it for submodules falsely penalizes them.

In most cases, any area definition is acceptable as long as the choice is clearly stated.

Completing the Efficiency Measurement

The solar cell to be measured is exposed to simulated sunlight. As a resistive load is varied from open-circuit through short-circuit conditions, the cell's current–voltage characteristics are measured. Figure 16 shows the current–voltage characteristics of one of the best cells ever measured at SERI. It is a cell made at SERI by Jerry Olson and is of a rather complex two-junction structure. By knowing the area of the cell (0.25 cm^2) and the power of the simulator (100 mW/cm^2), we know the amount of incident power on the cell itself (power per unit area times cell area; 25 mW). The current–voltage characteristics measured by the simulator give us the cell's power output (open-circuit voltage times short-circuit current times fill factor). We then divide the power-out (6.82 mW) by the power-in (25 mW) to get the cell's efficiency (27.3%) under standard conditions.

We may also do this measurement under nonstandard conditions—such as outdoors. It is natural to think that the outdoor efficiency would be a better measure than a conventional indoor one, but the opposite is the case. Outdoor conditions change not only according to location but by the atmospheric conditions at any given location. Efficiencies measured outdoors at noon on a clear day in one location can vary by about 20% from efficiencies in another location or from those measured under standard conditions indoors. Uncontrollable variations of this sort are completely unacceptable. Hence, the stan-

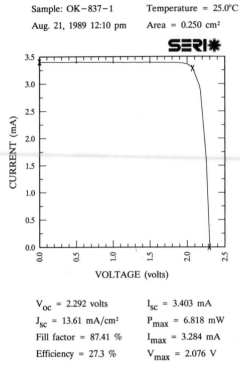

Sample: OK–837–1   Temperature = 25.0°C

Aug. 21, 1989 12:10 pm   Area = 0.250 cm²

V_{oc} = 2.292 volts	I_{sc} = 3.403 mA
J_{sc} = 13.61 mA/cm²	P_{max} = 6.818 mW
Fill factor = 87.41 %	I_{max} = 3.284 mA
Efficiency = 27.3 %	V_{max} = 2.076 V

Figure 16. The current–voltage characteristics of a high-efficiency cell made by Jerry Olson of SERI (GaInP/GaAs). Current–voltage curves are measured indoors using a solar simulator to mimic the distribution and intensity of the standard SERI global AM 1.5 spectrum. This cell is one of the best (most efficient) ever made.

dard indoor measurements are a much more reliable and fair measure of a solar cell's efficiency.

All of this is part of the never-ending battle to find the perfect measure of solar cell performance. But even the standard indoor measure of cell performance has its limits. Other measurements may be better for characterizing real, in-the-field energy output year-round at specific locations. For instance, standard measurements are done at 77° F (25° C), whereas cells actually perform at much higher temperatures (around 120° F) and thus have lower efficiencies. Yet actual, operational cell temperatures vary enough to make it difficult to choose any one to call a standard. In our cost calculations, we compensate for this temperature-dependent variation by subtracting 10% efficiency for real systems.

In fact, no perfect measure of PV performance can ever exist. How would anyone ever account for the variety of sunlight at different locations under different climates? Yet these variations might favor a cell that measured lower in efficiency under standard conditions. Over a year's time, such a cell might produce more electricity than one measured at higher efficiency under standard conditions indoors (although the opposite would be the expected norm).

Furthermore, standard PV efficiencies, by themselves, can also be misleading as far as the usability of the output of different cells. For one thing, electricity from a more efficient cell may cost much more than electricity from one that is less efficient. Electricity cost is a ratio of the cell's output to how much it costs to make it; the latter is hardly ever mentioned when efficiencies are given, but it can be overwhelmingly important.

We need efficiency measures (and knowledge of their limits) to integrate PV device performance into evaluations of PV systems. With them, we can include realistic circumstances such as higher temperatures, wiring losses, and module and BOS costs. When we also know how much a cell costs to manufacture, we can make proper comparisons of the cost-effectiveness of various PV options.

6 ☼ Silicon Cells

Conventional Processing

PV cells made of silicon have been the workhorse of the PV industry since their invention at Bell Labs in the 1950s. Use of silicon cells in space to power satellites gave silicon a technical lead that stood it in good stead for the next twenty years. Even now, silicon cells account for the major proportion of electricity production from PV (80% of US production). Their development and wide use has given PV a foothold in numerous markets and provided a record of outdoor reliability that would be the envy of most technologies.

Generic Silicon Cell Requirements

Silicon is an indirect band gap semiconductor. About 100 microns of silicon is needed to absorb about 95% of the photons in the solar spectrum. This has two very significant implications:

☐ Cells must be relatively *thick*.
☐ The silicon material must be of a *very high quality*. Most of the solar photons absorbed by silicon cells are outside of the 1-micron-thick built-in field region. To collect a large pro-

portion of the light-generated electrons and holes, silicon cells must depend on the diffusion of carriers to the distant field region.

This latter implies that silicon material must be very pure and it must be used in either single-crystal form or something very close to single-crystal. Both of these requirements—purity and structural perfection—add significant cost to the technology.

Silicon Feedstock

Silicon is one of the Earth's most abundant elements (20% of the Earth's crust). Yet in its most familiar form—sand—it is of little use for PV because of the presence of impurities. Instead, silicon for PV cells is refined from silicon dioxide (silica) taken from the ore quartzite. About a million metric tons (10^9 kg) of this so-called metallurgical-grade (MG) silicon is produced annually, mostly for the steel and aluminum industries. Although about 98% pure and costing only about $1/kg, it is far too dirty for use in PV cells or for making computer chips, the other major electronic use of silicon.

About 5000 metric tons (5×10^6 kg) of impure, MG silicon is refined further for electronics and computer applications. This so-called semiconductor-grade (SeG) silicon is purer and far more expensive than MG silicon. Impurities are reduced from the 2% level to a parts-per-billion (ppb) level. Cost goes from $1/kg to $20–$30/kg.

The conventional method for purifying silicon is called the Siemens process. Its main feature is a distillation step. MG silicon is chemically combined with hydrogen chloride to produce a liquid called trichlorosilane. This liquid is fractionally distilled and purified by multiple evaporation and condensation steps in much the same way that alcohol is distilled. Purified trichlorosilane is then heated and reacted with hydrogen to produce very pure silicon, with hydrogen chloride as a by-product.

The process is costly because the distillations are very energy-intensive, and they have a relatively low yield of purified silicon (under 30%).

Typical silicon cells require over 1 kg of silicon per m^2 of area. At current prices, such a layer would cost—by itself—a minimum of about $20/m^2$. Processing losses increase this cost.

Silicon costs have come down dramatically over the last 10 years, from over $100/kg in the early 1980s to their present levels. In the late 1970s, the cost of silicon all by itself was considered a severe roadblock to the technology. Today, silicon cost is less of an issue, although at $20/m² silicon cost still remains a concern. However, continued cost reductions are possible because (1) the conventional Siemens process can be modified to reduce its cost and (2) several innovative, alternative silicon processes are being developed.

One promising process that has reached a relatively mature stage was developed by Union Carbide Corporation with Jet Propulsion Laboratory funds from the Department of Energy. Like the Siemens process, the Union Carbide process is done in two steps. In the first step, the MG silicon is chemically changed to silane rather than trichlorosilane. Silane is a gas at room temperature. As such, it can be distilled at low temperatures, saving energy costs. After distillation, the pure silane is converted back to silicon via a process that is also designed to be inexpensive.

Another promising process was recently commercialized by Ethyl Corporation using a fluidized-bed approach to purifying silicon. This and other steady progress toward lower-cost silicon suggests that the present low cost of silicon, $20/kg, will hold or even drop some more.

After pure silicon has been obtained, the next step is to fabricate the cell in such a way that the silicon layers are nearly perfect crystals. High-quality silicon is needed for good silicon cell performance.

Cell Processing

The most critical step in the manufacture of silicon cells is the actual fabrication of the silicon layer from the purified silicon feedstock.

Over the years, a number of alternative methods of making silicon layers have been developed. Each has unique advantages and limits. To this day, none has emerged as the best method of making silicon in the sense that no approach dominates existing manufacture or plans for future manufacture. Each has its adherents. The four categories of silicon processing are:

1. Single-crystal boules (cylinders)
2. Ribbon growth
3. Cast material
3. Thin-film approaches

The growth of cylinders—or boules—of single-crystal silicon is one of the most frequent methods, even now, of making silicon cells. Slightly over 50% of US PV manufacturing falls in this category, much of it by one company—ARCO Solar (now Siemens Solar Industries) of Camarillo, California. In this chapter, we will focus on this particular baseline technology.

Single-Crystal Boules

The highest quality silicon and the best silicon cells result from the growth of boules of single-crystal silicon. The original work on this process of making high-quality silicon crystals dates back (1917) to the Polish physicist J. Czochralski, and is called CZ silicon.

The modern-day CZ process begins with SeG silicon feedstock. The feedstock is melted at high temperatures (over 1400° C) in a relatively nonreactive container called a crucible. The crucible's non-reactivity is an important criterion for making excellent crystals since at such high temperatures, molten silicon will dissolve or react with most materials. Such reactions would introduce impurities to the melt, degrading silicon quality and cell performance.

After many years of experimentation, the most commonly used crucible material is silica (silicon dioxide). This is also called fused quartz or simply quartz. Quartz has been a very successful crucible material, but it has a few problems. At 1400° C, it begins to soften, so it must be supported on all sides. More importantly, even quartz very slowly dissolves in molten silicon. This adds three kinds of impurities to the melt: unreacted quartz, oxygen, and other impurities already present in the quartz.

The next process step is called crystal pulling. A small (6–12 mm), single-crystal silicon seed crystal is touched to the molten silicon. As it is withdrawn, a raised meniscus of molten silicon forms between the seed and the melt. Both the seed and the crucible are oppositely rotated (10–40 rpm) to favor smooth, radial growth. At first a long, thin neck of single crystal is formed under the seed as the seed is pulled upward at about 30 cm/h. Then the rate of pull is dropped to about 2.5 cm/h and the diameter of the growing crystal increases to a desired size (say 10 cm), whereupon the rate of pull is adjusted to maintain this diameter.

During growth, liquid silicon is being depleted from the melt. To assure smooth growth, the height of the constantly depleted melt is adjusted to follow the location of the contact between the liquid and solid silicon. Also, care is taken to keep temperatures within the melt constant as the silicon melt is depleted.

To terminate growth, the pull rate is increased to allow the formation of a long, narrow neck at the bottom of the boule. Then the neck is withdrawn. The thicker boule itself could not be withdrawn rapidly because its sudden cooling would result in crystal defects. Typical lengths of completed boules can be 1–2 m.

Growth is accomplished in an inert atmosphere such as flowing argon. The argon reduces the incorporation of impurities from the other, more reactive ambient gases. If care is taken, the boule can be of the very highest-quality single crystal.

Dopants and Impurities

The key aspect of any PV cell is its built-in electric field. We have often stated that its formation and placement are controllable, but we have never stated how. The first step in making such a field in a silicon cell is the fabrication of material dominated by either electrons (n-type) or holes (p-type).

Silicon can be either an n-type or a p-type semiconductor. One way to make it n-type (dominated by electrons) is to somehow incorporate an element such as phosphorus into the silicon crystal lattice. Phosphorus has one more outer electron than silicon. If the phosphorus can be made to take a place in the silicon lattice, it will contribute that extra electron to the conduction band.

To make silicon p-type (dominated by holes), boron is incorporated in the lattice. Boron has one less outer electron than silicon. It will grab other electrons from localized positions in the lattice, creating an increase in free holes.

One of these impurities—usually boron—is introduced to the

silicon melt before growth is begun. Then as the melt solidifies, some fraction of the boron passes into the boule. As it does so, boron atoms take up places in the silicon crystal lattice, doping it p-type. Not all the boron in the molten silicon moves into the solidified boule. This is because the interface between molten and solid silicon tends to retard the movement of impurities from the melt. This has two effects, one bad and one good. The bad one is that the concentration of boron or other dopant will build up within the melt as more silicon solidifies. More boron will be available from the melt for incorporation—and will be incorporated—as the process proceeds. This will result in an undesirable dopant gradient in the boule (more at the bottom, last-to-grow section). Such a gradient would cause cells made from one part of the boule to have different properties than cells from another part. In practice, several subtle adjustments (e.g., varied stirring rates, varied growth rates, modified temperatures) are used to minimize dopant gradients. In fact, boron dopant is usually chosen for incorporation in the melt because it has less of a tendency to segregate than other dopants (e.g., phosphorus), so undesirable doping gradients are minimized.

The positive effect of the tendency for impurity segregation is that some fraction of other, unwanted impurities are also segregated within the melt. This improves the purity of the silicon, even beyond that of the SeG feedstock.

Wafering

To make cells, a boule of single-crystal silicon must be cut into many thin wafers. Cutting is accomplished by sawing a boule with multiple blades or abrasive wires. The thickness of the wafers is not determined as much by the needed thickness of silicon to absorb sunlight (about 100 microns) as it is by structural requirements. It is difficult to slice wafers thinner than 300 microns without wafer breakage; and if made too thin, Si wafers would be too brittle to handle in subsequent processing steps.

A serious drawback of wafering is that about half of the precious silicon is lost as sawdust. This material is part of a liquid slurry and is a total loss. Unless it is reprocessed—which may be done, but at a substantial cost—it cannot be returned to the melt.

After sawing, wafers are chemically etched and then cleaned. Chemical etching consists of bathing the wafer in an acid, which removes undesirable surface damage caused by wafering. Chemical etching of the wafer's surface removes a layer of damaged material. After cleaning (to remove the chemical residue), the wafers are ready for the next crucial step: forming the built-in field.

Junction Formation

A PV cell's built-in field is induced when a p-type region and an n-type region share a common boundary. The silicon wafers emerging from CZ growth can be either p- or n-type, depending on whether boron or phosphorus is introduced to the melt. As stated, boron is usually chosen, and typical wafers are p-type.

To make a p–n interface, we must make the top layer of the wafer n-type. This can be done by introducing a greater amount of n-type dopant (phosphorus) to that region. The n-type dopant is added in enough quantity to overwhelm the p-type dopant. Several methods of doing so have been developed, the most common one being a thermal in-diffusion of dopant from the surface of the wafer.

To do this, phosphorus is introduced by passing a gas such as phosphorus oxychloride over a heated wafer. This forms a phosphorus-containing oxide on all the surfaces of the wafer (front and back). Diffusion of phosphorus into the silicon occurs because the process is done at elevated temperature — 800–900° C. After about 20 minutes, enough phosphorus moves into the outer layers to overcome the presence of p-type dopant. A thin exposed layer all around the wafer becomes n-type.

The oxide surface layer is then removed by an acidic chemical etch. In addition, the unwanted n-type regions are removed from the bottom and sides of the wafer. An np interface remains at the top of the wafer. At this interface, a built-in field occurs because of the juxtaposition of n- and p-type material.

Many wafers can be treated with phosphorus simultaneously in a large chamber. However, various wafer handling steps subsequent to the dopant diffusion are required to etch off the unwanted n-type regions on the bottom and sides. To achieve low cost, these steps must be automated.

Completing the Cells

Two steps remain before cells are done: adding metal contacts on both sides of the wafer and reducing top-surface reflection with an antireflective coating.

Metal contacts can be deposited in a variety of ways. The conventional method is called vacuum evaporation. The metal is heated to evaporation and deposited along a line-of-sight path onto the silicon wafer. The wafer is kept at room temperature to minimize undesirable movement of the metal into the silicon, especially into the sensitive top surface. A grid is deposited either by evaporating the metal through a shadow mask (a mask with spaces corresponding to grid lines) or by depositing a uniform metal top layer and then etching off everything except thin grid lines. The back contact is formed in a subsequent step in which metal is deposited all across the back of the cell.

Unfortunately, various problems have caused a complex, three-layer metal sandwich to be used for contacting silicon. Titanium is deposited first because it has good adherence to silicon. Silver would be next (because it is extremely conductive); but in actual use, silver cannot prevent the corrosion of titanium in most environments. Instead, palladium is deposited between the silver and titanium to protect the titanium. The contacts are activated by heating the wafer to about 500–600° C. Drawbacks of this process are the expensive metals, the loss of a substantial fraction of the metals during vacuum evaporation, and the slow throughput of the process, which raises capital costs.

Since over 30% of sunlight would be reflected off the top of bare silicon cells, an antireflective coating is needed. A thin layer of a transparent, nonconducting material called a dielectric is deposited to reduce reflection to acceptable levels (under 10%).

Cells to Modules

Modules are the actual product of PV manufacturing. To form a module, enough crystalline silicon cells have to be connected together to produce meaningful power. Module sizes range from 1 to 5 square feet. Power outputs are in the 30 to 50 W range.

Cells can be 10–100 cm^2 in area. To connect cells, wires are soldered to the cell contacts and then interconnected. The cells are mounted in a frame and then sealed behind glass for long-term durability outdoors. This is called encapsulation. Somewhat costly handling steps—interconnecting and mounting individual cells—are implicit in this procedure. This is a major drawback of the silicon cell technology and is overcome or reduced in some of the other technological choices—e.g., in thin films and concentrators.

Status of the Baseline Silicon Technology

All PV approaches can be judged on the basis of three criteria:

□ Efficiency
□ Manufacturing cost
□ Reliability outdoors

Based on these (especially on efficiency and cost), a reasonable estimate of the cost to the consumer in cents/per kilowatt-hour can be made using the method outlined in Chapter 3.

Reliability is not much of an issue with the standard silicon technology. The efforts of many research groups and manufacturers over the years have resulted in a product that does not degrade. Module warranties of 10 years are common; actual useful life approaches 30 years. For the purpose of estimating costs, we shall assume a 30-year life. The basic stability of silicon, as well as the years of careful work in cell interconnection and module encapsulation, assure that this is a reasonable estimate.

Efficiency and Cost Status

Later on we will look at the best silicon cells. In this section, we are interested in understanding the costs of today's CZ silicon technology.

Before the oil embargo in 1973, silicon modules cost about $5000/m^2 and were only 5% efficient (50 W/m^2 at peak sunlight). This translates to over $7.50/kWh (assuming intermittent AC output with no storage). Compare this exorbitant cost with the cost of conventional electricity, i.e., about 5 *cents*/kWh.

By 1978, the cost of silicon modules had dropped to $1300/m², a fourfold decrease. Module efficiency had increased to 6–7%. AC electricity cost had dropped to *only* about $1.75/kWh.

In the early 1980s (1983), cost had dropped to about $700/m², and efficiencies of typical modules had reached almost 10%. PV electricity cost was down to 65 cents/kWh.

Now (1989) the cost of a typical CZ silicon module is about $500/ m², and efficiencies have reached the 13% range. AC electricity costs are in the 34 cents/kWh range. Figure 17 shows these trends in performance and cost.

Projecting from the existing trends, one might expect to see 16%- efficient modules costing $400/m². Assuming that we achieve the long-term balance-of-system goals (about $64/m², with power-conditioner), this would translate into AC electricity costs of about 16 cents/ kWh.

This is well above a 5 cents/kWh goal. One might argue that economies of scale would substantially reduce module costs if manufacturing of sufficient capacity were achieved. Similarly—as we will see later—impressive efficiency gains are being made in the labora-

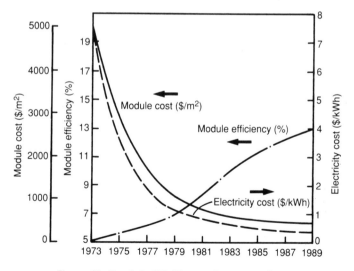

Figure 17. Trends in CZ silicon performance and cost.

tory on small-area cells, and these may yet push module efficiencies above those assumed here. If one uses an optimistic estimate (say 20% modules costing $250/m^2), the price of AC electricity (with no storage) would be about 9 cents/kWh. This remains on the higher end of the potential cost range of future PV technologies.

Conclusions

The baseline CZ silicon technology has made important contributions to the development of PV. It is a proven technology that is one of the best alternatives today for producing PV power. If major requirements for PV electricity were federally mandated on an emergency basis in the next 5 years, crystalline CZ silicon modules would almost assuredly occupy a large proportion of those produced *because they have a proven track record.*

Much of the motivation for continued research into CZ silicon and into related silicon alternatives emerges from the need to improve on the cost and performance of the existing CZ silicon approach. *The silicon technology is a crucial one to pursue because it provides a fully proven PV approach.* But newer approaches—either modifications of CZ or totally new ideas—have been, and should continue to be, investigated with the purpose of reaching even lower PV electricity costs.

7 Silicon Cells
Advanced Processing and Designs

In the past fifteen years, perhaps half a billion dollars has been spent on the research and demonstration of crystalline silicon cells. Much besides conventional CZ growth has been investigated. Some examples of advanced silicon processing are:

- [ ] Cast silicon
- [ ] Ribbon growth
- [ ] High-speed melt spinning
- [ ] Thin crystalline silicon

The advanced growth techniques tend to improve growth speed, make growth simpler and less costly, avoid sawdust losses, or reduce subsequent handling steps. Those that are most effective at reducing costs, however, tend to produce silicon of lower crystal quality than CZ silicon. Impurities may be higher and lattice perfection may be severely reduced. Rather than being single-crystal, the silicon may be polycrystalline—tightly packed grains of submillimeter sizes. Naturally, the presence of these grains degrades performance substantially in silicon cells, which are devices that are highly dependent on long diffusion lengths. Efficiencies in the silicon devices made by the most inexpensive processes substantially trail those made with the best quality material.

As with other PV technologies, there is a split among silicon researchers. Some emphasize reducing manufacturing costs, and others believe that reaching ultrahigh efficiency is the crucial goal. While new processing techniques have been developed to lower costs, much work has also been done on the other extreme: optimizing single-crystal silicon cell efficiencies. As a result, efficiencies that would have seemed pipe dreams at the beginning of the decade have now been realized. Along with this progress has come a set of new techniques for understanding all solar cells and increasing their efficiencies.

The progress of the 1980s has created an opportunity for the 1990s: synthesizing the impressive advances in low-cost manufacturing with similar advances in efficiency.

Cast Silicon

Perhaps the easiest method of making silicon, at least from a conceptual standpoint, is casting. In its simplest form, it is the solidification of molten silicon in a mold—usually in the shape of a large, rectangular ingot. Solidification is relatively simple but less controllable than CZ boule growth, and it always results in a polycrystalline structure. Also, precipitates from the mold cannot be segregated from the solidifying silicon as they would be in a CZ process. Thus, material quality is sacrificed for ease of fabrication and increased production rate.

The silicon ingot that emerges from casting is first sectioned into long, boxlike ingots, each of about 10 cm by 10 cm in cross section. These are then wafered in much the same way as CZ crystals. Sawdust losses are similar to those in CZ wafering. But in this case, rectangular rather than round wafers result from the sawing process. Rectangular cells have a tremendous advantage in terms of module efficiency, because they can be tightly packed. Circular cells pack poorly (22% is lost to the unused areas between circular cells), and modules made of them naturally have substantially lower efficiency than the cells from which they are made. The improved packing in modules of rectangular, cast cells helps to make up most of the difference between their efficiencies and the higher efficiencies of round, CZ cells.

Two companies are currently making PV products using cast

silicon. They are Solarex (a subsidiary of AMOCO) in the US and Wacker SILSO (a subsidiary of Heliotronic GmbH) of West Germany. Wacker has attained 16.6% efficiency on a 2 cm by 2 cm wafer. Module efficiencies (5600 cm^2) are about 12% for Solarex material. Solarex is one of the largest manufacturers of PV modules.

Ribbon Growth

Despite the fact that small-grained cast material can be made at a far higher rate than CZ silicon, it suffers from sawdust losses and efficiency limits. A very different approach is ribbon growth. Ribbon growth is the formation of wide, thin ribbons of silicon that are suitable for immediate processing into cells. No sawing step is needed because the ribbons have the appropriate thickness for making good cells.

Ribbons can be grown by a variety of methods that produce material of different qualities. Some methods can produce single-crystal silicon that is nearly as good as CZ. Others are more like the product of the casting approach, since they result in materials having numerous small grains. Like CZ, ribbon growth can be designed to repress the transfer of impurities at the boundary between the melt and the crystal, making the grown crystal purer than the feedstock.

Ribbons can be categorized as those that require a mold and those that do not. An example of the first category is one in which a seed crystal is touched to molten silicon through a narrow, nonwetting slot in a mold made of a temperature-resistant material like graphite. The solidifying silicon is then pulled upward through the shaping mold, forming a continuous ribbon.

A modification of this mold-shaping process, called EFG, has been developed by Mobil Solar, a subsidiary of Mobil Corporation. Mobil grows a "nonagon" ribbon, which is nine ribbons connected together at their edges to form one, nine-sided shape. The ribbons of silicon are pulled simultaneously from a melt through a nine-slotted mold, and lengths of 6 m have been attained. By pulling nine ribbons at once, growth rate is substantially increased: ribbon area of 160 cm^2/min can be pulled. The ribbons are cut apart prior to cell processing. Mobil has made cells by this method of 15% efficiency and has substantial plans to commercialize them during the 1990s. In

1989, they won an award for a demonstration project (200 kW$_p$) that will provide a testbed for their technology. Mobil Solar has stated that it will be in a position to sell their EFG modules for utility-scale projects during the 1990s at about 20 cents/kWh.

A variation of this technique has been developed by Hoxan Corporation of Japan. Instead of pulling the ribbon vertically, Hoxan pulls it horizontally. They place a slotted mold at the edge of the surface of molten silicon held in a quartz crucible. Molten silicon enters the mold and is pulled horizontally from it at rates of about 100 cm^2/min. Hoxan, which calls their process cast ribbon, has reported 11%-efficient cells made from their material.

One of the drawbacks of all approaches that depend on a mold is the incorporation of impurities from the mold. An example of a ribbon process that does not use a mold is dendritic web growth. Two long thin crystals of silicon, called dendrites, are held vertically, parallel to each other and touched to molten silicon in much the same way CZ growth is initiated with a seed crystal. If the dendrites are close enough together, molten silicon forms a meniscus between them. This is then slowly pulled from the melt, forming a thin sheet of high-quality material. Rejection of impurities can be as good as CZ, and there is no mold touching the ribbon. Westinghouse developed this approach with funding from the Department of Energy. Their ribbons are about 125 microns thick (much thinner than wafers), can be grown at a rate of 5 cm/min, and are about 6 cm in width (i.e., the areal growth rate is 30 cm^2/min). They have made 17%-efficient cells and modules that are 12.7% efficient (4000 cm^2). The advantages of their approach are: the production of high-quality, nearly single-crystal material from thin cells, i.e., with good silicon utilization. In 1989, Westinghouse said that they would expect their web silicon modules to cost about $0.75/W$_p$ in full production, which would translate to electricity at about 6.5 cents/kWh (if the BOS long-term goals were assumed). This would be a very important achievement.

A German company, Siemens Research Laboratories (Munich), has developed a high-rate ribbon process that has no mold. Called HSW for horizontal supported web (Figure 18), it is based on a technique of pulling a ribbon horizontally from the surface of molten silicon. A carbon mesh is used to begin growth of the silicon. The screen then forms the top surface of a 400 to 600-micron silicon sheet.

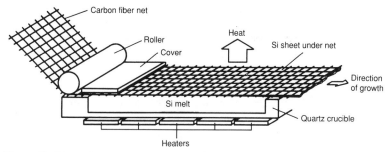

Figure 18. The horizontal supported web (HSW) silicon ribbon-pulling method developed by Siemens Research Laboratories in Munich.

Grain size is about 1 mm, and sheets as long as 7 m (6 cm wide) have been pulled at rate a of 1 m/min (1000 cm²/min). Cells of about 10% efficiency have been made by this high-rate process.

Another German company, Wacker Heliotronic of Burghausen, has developed a high-rate ribbon process called RAFT (ramp assisted foil technique; Figure 19). Molten silicon is solidified on a vertical ramp as the ramp is raised past the silicon melt. The ramp is specially

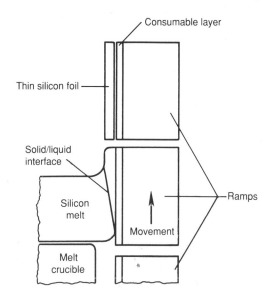

Figure 19. The ramp-assisted foil technique (RAFT) developed by Wacker Heliotronic of Burghausen, Germany.

coated with a sacrificial layer that is consumed during solidification, allowing the ramp to be coated again and reused after a silicon foil of 10 cm^2 area and 150–250 micron thickness is pulled free from the ramp. Extremely high rates of 10 m/min (10,000 cm^2/min) have been reached. Cells of 10% efficiency have been made on RAFT material. A different technique has been developed by the Fraunhofer Institute of Freiburg, Germany. Called SSP (silicon sheets from powder), the method uses silicon powder as a feedstock. This is placed on a supporting plate of quartz or silicon. The surface is melted and then removed, giving a free standing sheet. This sheet is then passed through a heat zone and the grains are melted and regrown in larger sizes. Cells of 12% efficiency have been reported. Although growth rates are low (about 2 cm/min) and sheets are still relatively small (10 cm by 15 cm), the method shows promise since it requires relatively inexpensive equipment.

Melt Spinning

An approach that could eventually achieve very high speed is melt spinning. In this case, molten silicon is poured onto a spinning disk which spreads the molten silicon outward into a narrow mold of the desired shape and thickness. Extremely high speeds can be attained but at the cost of crystal quality.

An example of this method is called spin casting, developed by Hoxan Corporation. They have achieved high growth rates of 100 cm^2/min. Poor crystal quality has kept them from raising this rate further. Sheets are about 500 microns thick. Solar cell efficiencies (100 cm^2) were reported to be as high as 11%.

Thin Silicon

It is possible to deposit relatively thin layers of silicon (under 50 microns) on ceramic substrates. This would reduce the amount of silicon used in silicon cells by about 75%. By doing so, silicon material costs could be minimized. However, the problem has been that cell efficiencies have suffered, and the ceramic substrate is not necessarily inexpensive. Since silicon is an indirect band gap material, much current would be lost in such thin designs *if other strategies were not adopted*. But methods of increasing current despite thin layers do exist. For instance, two light-trapping ideas are to texture the front surface (see Figure 22, below) to refract light (increasing relative path length) and to make the back surface reflective. The latter allows light that is not absorbed to reflect back into the cell. Theoretical studies actually suggest that thin silicon cells can have efficiencies as high as 19% if they are made correctly. A small company in Maryland, Astropower Inc., is a leader among those developing thin silicon. They have proposed several design rules for making efficient cells:

1. Grain size at least twice the thickness of the silicon thickness (e.g., for 50-micron silicon films, 1-mm-wide grains).
2. Diffusion lengths at least twice the film thickness (i.e., the same as the width of the grains).
3. Grain boundaries with a minimum of recombination centers.

Astropower has been successful in making 16%-efficient silicon cells of 100-micron thickness. Similarly, they have reached 10% efficiency on 100 cm^2 substrates. These latter are a new material that is reportedly of low cost. Several challenges remain:

□ The need to achieve the same or better results with thinner silicon layers.

□ A method of making very large areas—over 1000 cm^2—and interconnecting cells on those areas without incurring significant handling costs.

The idea of making thin silicon layers on inexpensive substrates is a good one that closely mimics the generic thin-film approach for achieving low-cost PV (see Chapter 8). Design procedures exist for using silicon successfully, at least in terms of making efficient cells. The ongoing effort to develop thin silicon is an attempt to exploit the strengths of silicon—the great knowledge we have of its properties, its abundance as a feedstock, and its benign environmental character. As we will see in the next chapter, thin films are a powerful path toward low-cost PV. Perhaps thin-film crystalline silicon on low-cost substrates will eventually take its place as one of the leading thin-film options.

Future of Low-Cost Processing of Silicon

All of the existing low-cost approaches suffer from some sort of cost/performance problem. Either rates are too low but quality is exceptional, or vice versa. Very low cost (near $70/m^2$) must eventually be achieved to compete with other developing PV options.

Progress in crystalline silicon is important because crystalline silicon PV is a baseline technology. It guarantees a method of making PV electricity at some reasonable cost. It is the standard against which others are measured. Continued progress in faster, better quality ribbon growth or in thin silicon will lower the cost of these technologies. At worst, these expected advances will put a reasonable upper limit on the cost of PV. On the other hand, progress in low-cost silicon may accelerate, and low-cost PV goals may be attained a lot sooner than anyone thinks.

Advances in High Efficiency

On the other end of the spectrum of silicon R&D are ongoing advances in very high-efficiency cells. Some maintain that advancing the efficiency of very high-quality silicon material and then reducing the cost of that material through process modifications is the way to lower costs. Scientists have made great progress over the last decade improving the efficiency of the best silicon cells. In fact, their work has pushed the theoretical limits higher by showing that via clever designs, cells with new conversion possibilities could be made. When the 1980s began, the expected theoretical efficiency limit of silicon was 22%. *Now that efficiency has actually been surpassed.* New limits have been calculated based on optimizing properties that the original studies considered unalterable. The new efficiency limits for silicon are in the 30% range. This is a clear example of the fact that PV has consistently outstripped expectations. Conversion efficiencies of sunlight to electricity with PV devices are not so much limited by theoretical considerations as they are by what is doable, technologically. The potentially more efficient PV structures that are waiting in the wings today seem almost out of reach to us from a technological standpoint.They probably appear as improbable to us as even the simplest PV device must have seemed to its pioneers in the 19th century. But given analogous technical progress in the next century, these very complex structures may become ordinary, and we may be performing research on PV devices that push the limits of conversion efficiency 50–60%, or higher.

Evolution of High-Efficiency Silicon Cells

Workers at Bell Labs were able to make 10%-efficient cells by 1957 using a wraparound cell structure. In this structure (Figure 20), a highly conductive p-type region wrapped nearly all the way around the cell. A key feature of the cell was that both contacts were made on the back of the cell, one to the p-type material, the other to the n-type. In this way, no shading of the top surface occurred, increasing current. This was a good idea and actually has resurfaced subsequently in several new guises. At the time, however, it was of limited usefulness because the path of carriers from the front surface to the contacts on the back led to very high series resistance. Lower voltages outweighed the benefit of high currents.

Figure 21 shows the classic PV cell that emerged when only the top region of the cell was p-type and a front metal grid was designed to replace the wraparound contact. The grid minimized series resistance (voltage) losses at the cost of slightly lower currents because of grid shading. Overall efficiency moved up to about 11–13% and stayed there throughout the 1960s.

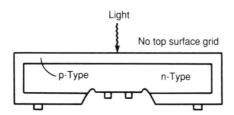

Figure 20. The wraparound cell was an early silicon cell design in which shadowing was avoided by putting both contacts on the back. Series resistance losses in the p-type region were large, and the design was abandoned for over a decade until newer methods were developed for achieving the same effect.

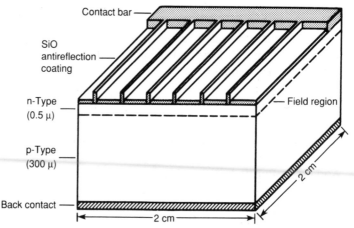

Figure 21. The conventional silicon solar cell design emerged early in the 1960s for space applications.

Surface Passivation

The next major step was conceptually different: a passivating layer was added to the back of the cell so that voltage was increased. The improvement was a result of a passivation technique that was in many ways analogous to the passivation of grain boundaries. At any silicon surface, there are more defects than within the lattice. Electrons at the surface, which would normally be sharing bonds between silicon atoms, are without neighbors with which to form those bonds. This produces an electronic defect called a dangling bond. Surfaces have many dangling bonds, whoch provide opportunities for free minority carriers (those that need time to diffuse to the built-in field) to recombine.

The same effect—undesirable recombination—also takes place near contacts between metal and silicon. The explanation for this is a little more complex. Contacts are most effective when there is a very highly conductive region in the silicon adjacent to the metal. But the very high doping needed to make these contacts causes a layer with many defects. The defects come from the extra, inactive dopants that are inadvertently added during the doping process. As noted previously, some added impurities can lodge between atoms in the lattice (interstitials). These defects provide numerous paths for recombination. Recombination can be reduced by passivation of defects. The first use of defect passivation was actually accidental. It was found that adding aluminum doping to the back of silicon cells raised their efficiencies. Several incorrect explanations were offered for this effect before it was found that it was caused by the passivation of the back surface. In this case, aluminum is a good p-type dopant of silicon. Adding a large amount near the back surface produced a p–p+ region (p+ meaning more p-type carriers). This had an effect analogous to juxtaposing an n- and a p-type region, i.e., it produced a slight built-in field between the differently doped p and p+ regions. The resulting electric field pushed electrons (minority carriers in p-type silicon) away from the back contact. This meant that they had less chance to recombine near the contact. Reducing this kind of recombination improved voltage (and to a lesser extent, current), and efficiencies increased in these unintentially passivated cells.

In the early 1970s, a major advance in efficiency occurred with the design of the "violet" cell (so named because of its violet appearance). Response to high-energy light (as measured by quantum efficiency analysis) had been poor in cells prior to the innovations of the violet cell. It was found that a dead layer existed on the surface of the silicon cell. This dead layer was relatively unresponsive in the sense that almost all the light-generated carriers formed within it were lost to recombination. As we know, high-energy light is the most strongly absorbed, and this was the light being absorbed (and lost) in the top surface dead layer.

On further analysis, the dead layer turned out to be the result of the thermal diffusion step that was used to form the n-type top layer of the cell. n-type dopant (phosphorus) was coated on the cell surface and then diffused into the cell by heating to 1000° C. This produced a gradient of phosphorus dopant that provided a good electric field with the p-type base region. However, near the surface, a very large amount of phosphorus remained in the form of silicon phosphide precipitates. These undesirable impurities were the source of the recombination.

By chemically removing the dead layer prior to putting on the top metal grid, a cell with a better high-energy response was made. In the meantime, a better antireflection coating was found that added to the amount of light entering the cell. Some adjustments were also made in the conductivity of the p-type layer based on a developing understanding of the details of how these devices behaved. As a result, an increase in efficiency occurred to over 15%.

The next major improvement was also related to reduced surface reflection. Scientists found in the mid-1970s that several chemical etches—hydrazine and sodium hydroxide, for example—could be used to produce many small pyramidal shapes (Figure 22) on the top surface of silicon cells. This was called texturing the top surface. The resulting cell was even less reflective than the violet cell and was called the black cell. The pyramids acted like an antireflection layer in the sense that light that was reflected had a good chance of striking— and being absorbed—in another face rather than being lost. For instance, if the probability of light being reflected once was 20%, its

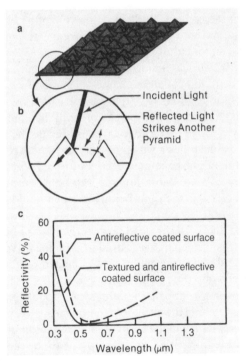

Figure 22. Texturing the surface of a silicon cell exposes pyramidal surfaces (a). Light that is nearly perpendicular can either be absorbed or reflected (b). If it is not absorbed (b), it will strike the surface of another pyramid and have another chance to be absorbed. Using both an antireflection coating and texturing (c) can reduce reflection losses to almost zero.

probability of being reflected twice was only 4%. Most light would get at least two chances to be asborbed, so almost 96% would be absorbed given the texturing.

Another advantage of the textured cell was to refract the light entering the cell. This meant that in effect the light had to pass through a thicker layer of silicon than it would if it were going into the silicon perpendicular to its surface. The result was to increase the effective absorption of the silicon. Low-energy light, which would normally pass quite deeply into the silicon, was absorbed closer to the built-in field. It had a better chance of contributing to the current.

Response of the black cell to sunlight was much improved over any of its predecessors. Efficiencies as high as 17% were demonstrated. This efficiency mark stood for almost a full decade before new improvements—based on a more sophisticated understanding of silicon cells—led to renewed progress.

Passivation in Earnest

Efficiencies were boosted in the mid-1980s when scientists were able to find new ways to apply passivating layers. In the microelectronics industry, silicon dioxide (SiO_2) has been used very effectively to passivate silicon surfaces. However, thick layers of SiO_2 are rather reflective when applied to silicon. In the early 1980s, scientists were able to find ways to use very thin (20 angstroms), nonreflective SiO_2 as passivating layers on the top surface of silicon cells. They applied SiO_2 over the entire top surface, including beneath the metal contacts. The SiO_2 tied up the dangling bonds at and near the silicon's surface. This significantly reduced recombination at the top surface and under the contacts. Cell efficiencies close to 19% were reached. A further improvement to over 19% was made when very small holes were left in the SiO_2 just beneath the metal contacts. This allowed better current flow without much increase in recombination.

A group led by Martin Green at the University of New South Wales produced the first 20% cell in 1985 using this approach and a new kind of surface texturing. Parallel triangular grooves were etched in the surface, allowing for current enhancement in comparison to surfaces made up of random pyramids. Subsequently (1989), Green's team made the world's most efficient silicon cell (23.2%) by further optimizing passivation layers and light-trapping methods.

The Point Contact Cell

Another advance was the point contact cell, developed by a group led by R. M. Swanson of Stanford University. It incorporated a unique design. Like the first silicon cells, it put both contacts on the rear of the cell, leaving the top surface fully exposed. However, the point contact cell departed in a radical way from existing designs. Not only were the contacts on the back, but the n- and p-type regions were also dispersed over the back.

The main portion of the cell was made using highly resistive silicon of the highest quality. We might call this material intrinsic, or i-type, after the fact that it was undoped and thus not dominated by a majority of either carrier type. Because it was of extremely high quality, it had very long diffusion lengths.

On the back of this intrinsic wafer, a checkerboard pattern of small, isolated n- and p-type patches was diffused into the silicon. Metal contacts were then applied to these patches.

A textured surface was etched on the top of the cell and a thin passivating silicon oxide was grown. Light entering the cell was absorbed somewhere in the intrinsic region. Because of the material's long diffusion length, most light-generated carriers were able to diffuse to the n- and p-type patches on the back, where electrons and holes were separated and contributed to external current flow.

These cells were actually designed for concentrator applications, which we cover in Chapter 12. Concentrators are PV systems based on focusing concentrated sunlight on small, highly efficient cells. For a period in the late 1980s, the Swanson cell held the record efficiency (28.2%) for any solar cell under focused sunlight. Their performance under normal sunlight was also exceptional. A cell as large as 8.5 cm^2 was measured at 22.3% in 1989.

Microgrooved Cell

The group at the University of New South Wales led by Martin Green made another important contribution to progress in silicon cell technology with a practical, high-efficiency contacting method called microgrooves. A narrow, lightly n-type layer was diffused on a p-type base layer as in many other high-efficiency approaches. Then the top layer was textured and then passivated with a very thin oxide layer, after which effective antireflection layers were also deposited. Long, thin microgrooves were made by cutting narrow (20 micron), 100-micron-deep grooves in the top surface of the cell with a laser. The laser cut through all the top layers, including the electric field region. The next step was to repair the field region by applying more n-type dopant across the top surface and heating the cell in a furnace. This heat treatment had little effect on most of the surface (which was protected by the oxide and antireflection coatings), but within the grooves it reestablished the np region. Then a metal was plated selectively into the grooves, forming a contact. The contact was especially effective because it had a large, vertical area of contact with the silicon, but its narrow cross section meant that it did not block much light (only 3%). Cells were fabricated with efficiencies of 19%.

Although this cell was a little less efficient than their 23% cell (referred to above), it was more practical in the sense that it could be more easily manufactured. Green has stated that the adoption of various new, high-efficiency techniques would allow for the production of silicon modules of 20% efficiency. However, the synthesis of high-efficiency and low-cost manufacturing remains the crucial challenge for the silicon cell technology.

8 ☼ Thin Films

Thin-film solar cells—the use of ultrathin layers—are being developed with the aim of lowering the cost of PV manufacturing by close to an order of magnitude in relation to CZ silicon. The main features of this projected cost reduction are:

- [ ] Less material, therefore lower material costs
- [ ] Thinner layers, therefore faster processes and lower capital costs
- [ ] Processing of larger-area PV devices, thus reduced handling costs

The various thin-film PV technologies, based on materials such as amorphous silicon, cadmium telluride, gallium arsenide, and copper indium diselenide, attempt to take advantage of all these potentially large cost savings.

Lower Materials Costs

How much cheaper are thin films than the classic "thick film" PV material, crystalline silicon? Existing crystalline silicon modules require at least 1 kg of silicon per m^2 (including waste), which at today's low prices for silicon feedstock implies a cost of at least $20/m^2$ for silicon material. The cost of a thin film in a thin-film module depends on (1) the cost of the material of which the module is made, (2) the thickness of the material within the module, and (3) the utilization rate of the material during deposition. The latter is the fraction of the material that ends up in the film versus the material used in the process.

Depending on the density of a material, there are about 3–6 grams in a layer 1 m² in area and 1 micron thick. (This size is useful for calculations since it gives us the amount of material in a module a square meter in area and 1 micron in thickness.) But some of the typical thin films (e.g., $CuInSe_2$, CdTe, GaAs) contain feedstock materials that cost more than refined silicon. The most extreme of these is indium, at about \$300/kg; gallium is almost as expensive (\$200/kg); tellurium is also expensive, costing about \$130/kg. These are by far the most expensive materials being used for making thin films. Other materials (Cu, Cd, Zn, S, Se) are generally much cheaper—closer to \$1–\$10/kg. In calculating the cost of material, the cost for these inexpensive materials tend to be so small (pennies per square meter) that they disappear in the uncertainty of other costs.

Let's examine the worst-case example, indium. This will give us a ceiling for thin-film materials cost. The indium in copper indium diselenide is about a third of the material by weight. Copper indium diselenide has a density of about 6 grams/cm³. Since a cubic centimeter is equal to a layer 1 micron thick and 1 m² in area, there are about 2 grams/m² of indium in a micron-thick layer of $CuInSe_2$. The thickness of the copper indium diselenide in typical $CuInSe_2$ cells is about 2 microns. Thus, the amount of indium in the cells is about 4 grams/m². If one assumes a 70% material utilization during the process that made the layer, starting material of about 6 grams of indium would be needed to make a 2-micron layer of $CuInSe_2$. Thus, the indium in copper indium diselenide would cost about \$2/m² (6 grams × \$0.30/g).

The other materials in the $CuInSe_2$ cell are copper, selenium, cadmium, sulfur, zinc, oxygen, and molybdenum. Each is at least ten times cheaper than indium; most are a hundred times cheaper. Their total cost does not even match the cost of indium. So a high-end estimate of the cost of materials for thin films might be about \$4/m². This cost is less than a fifth of the cost of silicon in existing crystalline silicon cells.

The cost of materials for thin films would actually be higher, and more similar to the costs of crystalline silicon, were it not for the fact that materials used in thin-film cells do not have the same purity requirements as crystalline silicon. Crystalline silicon devices are far more dependent on diffusion lengths than are thin films, where most of the current is collected in or very near the electric field. Achieving the best material perfection—including high purity—is critical to successful crystalline silicon devices; for thin films, material perfection is much less critical. Thus, the purities of most feedstocks for thin films are no greater than 0.9999; for crystalline silicon, purity is closer to 0.99999999999999. Much of the cost of silicon results from having to reach extreme purity. In sharp contrast, thin-film cells—because they collect most of their current from drift rather than diffusion—do not need anything like silicon's required purity.

Because they require such small quantities of material, thin films can be made from relatively rare and costly materials that would otherwise be impossible to use (indium, tellurium, gallium). Many very interesting and potentially important thin films have been developed based on these novel materials. They form the basis for much of the progress in thin films. The fact that new, possibly more favorable semiconductors can be used in thin films is itself almost as important as the more obvious fact that the thin-film approach saves money through reducing materials use.

Lower Processing Costs

Thin-film cells have thinner layers. This provides an opportunity for reduced processing costs. The idea is that processes could be developed where very thin layers could be rapidly deposited on low-cost substrates such as glass. Examples of such processes exist. For instance, semiconductor materials suspended in solution have been sprayed on moving glass substrates with areal velocities of square meters per minute, far in excess of those of even high-speed silicon ribbon growth techniques. The cost of a module manufacturing process is proportional to the cost of the equipment divided by the speed with which the equipment makes layers. Low cost can be attained by using cheap equipment that can make large areas quickly. Intermediate costs are achieved by the combination of either low speed with cheap equipment; or high speed with expensive equipment. Even thin films can be expensive if they are manufactured using costly equipment incapable of rapid areal deposition rates.

Unfortunately, low-capital-cost, high-speed methods are the exception rather than the rule. As speed or simplicity increase, the quality of processed material can drop substantially. So many companies have been forced to opt for more expensive processes because they have not had the resources to develop truly low-cost processes that make high-quality materials. But truly low-cost thin films can only be attained when every process in the production line is chosen for simplicity, speed, high material utilization, and high yield. Without these conditions, the manufacturer may end up making a thin film that costs nearly as much as any other PV option. On the other hand, with ambitious cost optimization, cost reductions due to faster processing using inexpensive equipment can lead to significant reductions in processing costs versus conventional crystalline silicon.

Lower Handling Costs

Thin films are not thick enough to be self-supporting. They are deposited on substrates such as glass or metal foil, which then act as structural supports for the final modules. A subtle but very marked advantage of thin films is the possibility of making them on very large substrates. The largest crystalline silicon cells are usually about 100 cm² in size. But thin films have been made on glass sheets 100 times larger, and more. To make a module from crystalline silicon cells, the cells must be wired together, which is a clumsy, time-consuming operation. On the other hand, thin-film modules of very large sizes are made during the deposition of the thin layers on large substrates. Very many time-consuming and costly cell handling steps are avoided. In addition, because very large modules can be made (eventually perhaps several square meters in size), wiring costs associated with assembling modules into arrays can be much lower. These factors combine for major cost savings as well as increased operational simplicity and yield.

All in all, the key advantages of thin films—reduced material costs, less expensive processing, simpler handling—allow for very significant cost savings over crystalline silicon cells in their conventional "thick film" design.

Problems

Serious problems also exist, which have slowed the development of thin films substantially. A generic problem of all low-cost thin films is relatively poor efficiency. Thin films are made by depositing the key layers on an inexpensive supporting substrate such as glass. This produces rather poor-quality layers, because grown materials tend to mimic the quality of materials upon which they are deposited. For instance, silicon layers carefully deposited on single-crystal CZ silicon can come out as single-crystal layers. But layers deposited on less well-ordered (and much less expensive) materials such as glass or metal tend to produce very poor layers.

The reason is that glass or metals are not single crystals. Layers grown on them have imperfect lattices and are either polycrystalline or amorphous materials. Polycrystalline layers are made of many small crystallites or grains (about 1 micron in size) packed together. The grains may have lattice defects within them, and the boundaries between grains are also potential defect sites. Amorphous materials have even less structure than polycrystalline materials. The lattice of an amorphous material is random. Instead of a building with many exactly replicated rooms, an amorphous material's atomic structure would resemble a building made of Silly Putty—totally deformed. Most of the connections between atoms would be the same in the sense that one atom might still be connected to four others, but the angle at which they are connected would vary randomly. Amorphous materials also have boundary regions (like grain boundaries) where different amorphous layers grow together. Like grain boundaries in polycrystalline materials, they are sites of increased recombination. The behavior of amorphous materials is far different from that of crystalline ones, and usually much worse from the standpoint of PV requirements such as diffusion lengths or defect concentrations.

Low-cost substrates (rather than costly single-crystal ones) are absolutely essential for achieving low-cost thin films. Unfortunately, this same requirement is a severe burden since it means that cells must be grown on polycrystalline or amorphous materials. The result is poor-quality films. But the effort to develop successful thin films has consistently emphasized reaching *reasonable* efficiencies (10–15%) despite the poor PV properties of materials made by low-cost processes on inexpensive substrates. Fortunately, clever cell designs have been developed to minimize the negative effects of mediocre semiconductor layers.

Choosing a Thin Film

Successfully developing new thin films has been an extremely challenging effort. Materials suitable for thin films have had properties far different from those of crystalline silicon. In some cases, new thin films have been based on semiconductor materials previously unused or even unknown. These new semiconductors have been synthesized and developed exclusively for PV. Totally new processes for making these materials cheaply over very large areas have also been developed. The development of each of these new thin films has been somewhat equivalent to the development of a key technology such as the laser disk or the megabyte computer chip.

Very little in the evolution of PV cells for space—the driving force in PV development from 1957 to 1973—prepared us for the development of thin films, because space cells were optimized for high efficiency, not low cost. With cost as a major driving force, the terrestrial PV effort set off in new directions, i.e., thin films. To get started, potentially successful thin-film materials had to be selected.

The most critical quality of a potentially successful thin-film semiconductor is very strong light absorption. Thin films must be able to absorb almost all sunlight in less than a micron of thickness, otherwise thin layers cannot be utilized. Strong light absorption also allows them to get nearly optimal currents (at least potentially), because so much light can be absorbed in or next to the built-in electric field.

A band gap between 1 and 1.8 eV is suitable for maximum potential efficiency. This is based on calculations that show that these band gaps are appropriate for devices with theoretical efficiencies over 20%. Numerous compound semiconductors exist with suitable band gaps.

Another quality of value in a thin-film cell is a tendency to have a minimum of defects, either in the bulk of the material or along grain boundaries. All thin films have defects, but some materials seem to have fewer of them or to be less dominated by them. For example, dangling bonds at grain boundaries can be very powerful recombination centers in some materials. In others—those of greater value as thin films—grain boundaries have surprisingly few recombination centers. Since recombination tends to seriously affect voltage, the choice of materials having fewer or less-powerful defects is imperative. Finally, it is important to choose a material that can be made easily and cheaply.

The First Thin Film: Copper Sulfide

Even though thin films did not receive focused attention before the energy crisis, the first thin film dates back to the beginning of practical PV in the mid-1950s. Its discovery was accidental. Copper sulfide—the first thin film—has a colorful history that is illustrative of many of the quirks and qualities of thin films in general.

The first copper sulfide devices were accidentally made in 1954. Scientists observed that a PV response could be produced when copper contacts were electroplated on a semiconductor called cadmium sulfide (CdS). Electroplating is a method by which an electric current flowing through a solution causes a material (in this case, copper) to deposit on one of the electric contacts (called electrodes) through which the current is entering the solution. It was not known then, but when copper is plated onto cadmium sulfide, another semiconductor—copper sulfide (Cu_2S)—forms between the copper-containing electrode and the sulfur-containing cadmium sulfide. Under such circumstances, some of the cadmium in the cadmium sulfide is replaced by copper to form the copper sulfide compound during the copper deposition. An electric field forms between the two semiconductors because cadmium sulfide is always n-type and copper sulfide is always p-type. Thus, the simple deposition of copper onto CdS resulted in a thin-film cell because a built-in electric field was created at the interface between the n-type CdS and the p-type copper sulfide.

The CdS/Cu_2S device is called a heterojunction structure. Cells in which the electric field is formed at the interface of oppositely doped material of the same composition, e.g., n-type silicon and p-type silicon, are called *homo*junctions (homo for being the same); cells where the junction is between two dissimilar materials, e.g., n-CdS and p-Cu_2S, are called *hetero*junctions.

The simplicity with which CdS/Cu$_2$S cells can be made has always been the strongest point of the technology. But confusion existed for ten years over the actual cause of the PV effect in the devices. When the PV effect was first observed in these cells during the 1950s, the formation of copper sulfide under the copper electrodes was unknown. Instead, researchers attributed the effect to an electric field forming between the copper metal and the CdS; or to a p–n junction occurring in the CdS from diffusion of Cu into the CdS. It was only in the mid-1960s that researchers hypothesized that copper sulfide was being formed during the copper deposition on CdS. In fact, to this day, this cell is most frequently called the "cadmium sulfide" cell because the truly essential material, copper sulfide, is still habitually overlooked.

The decade of confusion about the true nature of the copper sulfide/cadmium sulfide cell is not atypical in the development of thin films, which frequently are based on wholly novel semiconductors with many subtleties. That the basis of the PV effect in these devices was unknown for ten years was not as much a failing of the researchers who were investigating copper sulfide as it was a reflection of the complexities involved.

Very Low Cost

As stated previously, the most attractive feature of copper sulfide was the extremely simple way that it could be fabricated. Not only was it first made accidentally when copper contacts were electroplated onto CdS; it was subsequently developed further using a very similar, simple method of dipping CdS *for about 5 to 30 seconds* in a chemical bath containing copper so that copper sulfide formed. In a sense, this kind of process—dipping in a liquid—is the kind of process that should dominate low-cost thin-film processing. Processes should be so cheap that any cost reduction would be almost inconceivable. *In the end, only these kind of processes will allow thin films to reach their true potential.*

A practical CdS/Cu$_2$S cell design is shown in Figure 23. This device illustrates most of the differences between thin-film cells and the more familiar crystalline silicon cells. Crystalline silicon cell processing starts with raw silicon which is regrown as a boule or ribbon and then fashioned into the proper size. With thin-film Cu_2S, glass is the starting substrate material. The pictured cell was made by a company, Photon Power (El Paso, TX), which brought copper sulfide close to commercial production in the late 1970s. The first layer that was deposited was a very highly conductive, high-band-gap (3.3 eV) material called tin oxide. This so-called transparent, conductive oxide acted as a top contact to the cell without obscuring any incoming sunlight. Such transparent, yet conductive materials are very important in the development of thin films; they allow good contact to the entire top of the cell without shading it by an opaque material such as a metal.

The method of depositing the tin oxide was almost as simple as the method of making Cu_2S: it was sprayed onto a moving glass substrate using a liquid containing tin. During deposition, the tin reacted with oxygen from the air to form tin oxide.

The next layer was CdS, and this was also deposited onto a moving substrate by spraying. Then the cell was dipped in a solution containing copper ions, which reacted with the CdS to form a very thin (0.1–0.3 micron), but highly light-absorbing layer of Cu_2S. To complete the cell, a metal was deposited on the back as a contact.

The founder of Photon Power, John Jordan, is a key figure in low-cost PV. He is still active today and doing pioneering work with another thin film, cadmium telluride (Chapter 11).

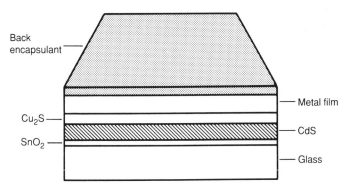

Figure 23. Schematic of a CdS/Cu_2S cell made by Photon Power, El Paso, Texas (c. 1980). Light comes through the glass (pictured on the bottom) to reach the Cu_2S absorber material.

Module Design and Pilot Plant Production

Copper sulfide was the first thin film to be brought into production. Although it has never been successfully commercialized, much progress was made that has proved useful to the development of other thin films. Innovations included the initial design of the large-area modules that have since become characteristic of thin films, as well as the design and development of the production lines needed to make modules in sufficient volume and least expense.

Two US companies attempted to commercialize copper sulfide in the late 1970s and early 1980s: Photon Power and SES (a subsidiary of Shell Oil). Photon Power was able to make large-area modules. By doing this, they demonstrated one of the key characteristics of thin films: the ability to make very large devices on a large substrate. This showed that the many handling steps that are needed in assembling a crystalline silicon module could be avoided in thin films.

Large-area thin-film modules are made up of many long, thin cells connected together. But when modules are made, the entire substrate is covered with a continuous sheet of semiconductor material. This continuous sheet must be subdivided into many long, thin cells to maximize module efficiency. To do so, automated processes have been developed based on making long, thin cuts, or scribes, with lasers or fine mechanical styluses. The procedure is a rather complicated one involving several sequential scribes and depositions. But the principle involved is that the top contact of one cell can be attached to the bottom contact of its neighbor. Current then flows between cells. In fact, the resulting current of the entire module is the current of only one cell (since all current is flowing through all cells simultaneously). But the resulting voltage of the module is the voltage of all the cells added together. This makes for high-voltage, low-current modules, which minimizes resistance losses (which rise as the square of the current). The individual cells of a thin-film module are usually made as long, narrow strips that run the length of a module. Each cell is about 1 cm wide.

Work in Cu_2S pioneered the development of these large-area, low-cost modules, which have become the model for all thin films. The scribing steps have been done by either lasers or with mechanical scribes.

The copper sulfide technology reached pilot plant manufacturing. Figure 24 shows a schematic of the production line at Photon Power (now Photon Energy). Glass was the starting material, and the finished modules emerged from the end of the line. Through the development of copper sulfide, many of the good theoretical ideas about low-cost thin-film production were demonstrated. Unfortunately, two of the problems with thin films—low efficiency and instability—were visible too, with devastating results.

Inefficiency and Instability

The best copper sulfide cell ever made was fabricated at the University of Delaware's Institute of Energy Conversion (IEC) in 1979. Measured then at 10.2%, it was the first thin film ever to achieve double-digit efficiencies. (Standards for efficiency measurement have changed and are now based on a different spectrum. The IEC cell would probably be in the 9% range if measured under the new spectrum.) In production, modules were much lower: 5% in efficiency. Progress might have continued to improve this efficiency to a level

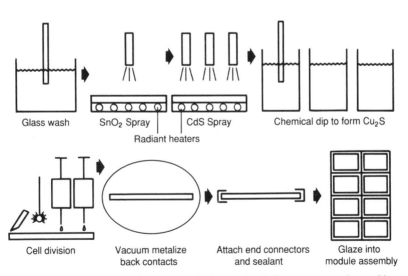

Glass wash   SnO$_2$ Spray   CdS Spray   Chemical dip to form Cu$_2$S
Radiant heaters

Cell division   Vacuum metalize   Attach end connectors   Glaze into
back contacts   and sealant   module assembly

Figure 24. Photon Power (El Paso, Texas) had a very simple, low-cost approach to making thin-film copper sulfide modules based on dip-coating copper sulfide on sprayed cadmium sulfide.

sufficient to be of use, but another problem became too severe to ignore: instability.

One does not normally expect a semiconductor to be unstable. Silicon is not. Under controlled circumstances (i.e., without water vapor attacking their metal electrical contacts), silicon cells could operate indefinitely. But copper sulfide is a material in which very small thermal or electronic forces can produce large changes. The root of the problem is that the copper in the copper sulfide lattice is relatively mobile. The bond holding it in place is weak. Under thermal or electronic stress (such as the presence of an electric field), copper moves in copper sulfide, radically changing the properties of a copper sulfide device. With sufficient stress, copper can actually precipitate from the lattice as solid metallic nodules. Formed within the electric field region, these highly conductive shunts between the top and bottom contact quickly ruin the device.

The copper sulfide modules were also very sensitive to water vapor and required very extensive encapsulation. In fact, by the early 1980s, a battery of degradation mechanisms had been identified in copper sulfide devices. Even in the best of circumstances, a module that had been optimally made for good performance would quickly deteriorate. Despite a decade of work aimed at stabilizing copper sulfide, the inherent mobility of copper in the lattice proved too difficult to overcome.

Despite the very low-cost approaches used to make copper sulfide, progress ceased because efficiencies were disappointing (below 10%) and stability became a severe and, finally, a fatal issue. Today almost no one considers copper sulfide an important thin film for the future, yet more recent thin films have benefited from its accomplishments.

Evolution of Other Candidate Thin Films: SERI's Role

Although copper sulfide was generally abandoned in the early 1980s, the evolution of other thin films had already begun. Among those aiding this evolution has been the Solar Energy Research Institute. Formed in 1978, SERI is a laboratory of the Department of

Energy with the mission of developing many forms of solar energy — biofuels, solar thermal systems, wind, and PV. Since its inception, SERI PV has specialized in thin films and other innovative PV options. Although SERI's overall budget fell by half during the 1980s (from over $100 million in 1980 to about $50 million annually today), its PV budget from DOE has been rather steady, in the $20-$30 million range during the 1980s. With that money, SERI has funded both in-house research as well as research taking place throughout the US.

Much of SERI's PV funds have gone into the development of thin films such as copper indium diselenide, amorphous silicon, gallium arsenide, and cadmium telluride. In some cases, SERI's funds have provided the majority of research money spent to support these technologies. Even in cases where US research has been privately funded, SERI's research money has frequently been essential to progress due to the fragile condition of the PV industry. Most PV companies have not made money. They have had little money for R&D. For many of them, SERI provided the small, but critical R&D money with which they built their technical base. Sometimes SERI's money (the US taxpayer's money) has been the difference between a viable PV technology and a dormant one. To that extent, an examination of the progress of PV (and of thin films in particular) is an examination of the potential capability of the US government in intervening successfully in favor of the development of a new high technology. PV could be seen as a test case for other, future, assistance to private development of technologies such as superconductors, high-definition television, and supersonic planes.

In the 1980s, when US industry was turning inward to maximize short-term profits, and when energy issues were being ignored because oil prices were sinking, it was fortunate for PV that a modicum of federal support was available. In terms of actual technical progress, the richness of the PV field and the commitment of its researchers did the rest.

In the early 1980s, SERI supported R&D to investigate a multiplicity of new thin-film materials. As the copper sulfide technology lapsed, three other thin films were already making rapid progress: amorphous silicon, copper indium diselenide, and cadmium telluride. Research interest shifted from copper sulfide in the late 1970s and early 1980s to these three materials. Subsequently, each has made substantial progress, and much of our optimism about the future of PV is based on expectations of their ongoing success. Through the development of these materials, the true potential of thin films to drop PV costs into an affordable range should be realized.

9 ☀ Amorphous Silicon

During much of the 1980s, the major focus of thin-film efforts was on amorphous silicon (a-Si). About $60 million of federal funds were expended, mostly through SERI programs. Meanwhile, global investment in a-Si was about half a billion dollars, some of it for non-PV uses of a-Si such as thin-film transistors and optically sensitive coatings on xerography drums. The potential value of a-Si for non-PV electronics is one of the major sources of its vitality: researchers from non-PV disciplines have helped broaden our understanding of a-Si.

As a material, a-Si is a disordered (i.e., amorphous) cousin of crystalline silicon. Although each silicon atom in a-Si usually has four nearest neighbors—just as it does in crystalline silicon—the bond angles connecting it to those neighbors are unequal. Instead of the 109° angles in crystalline silicon, bond angles in a-Si vary over a wide range. No orderly lattice structure exists, changing fundamentally the properties of the material. Instead of an indirect band gap of 1.1 eV (crystalline silicon), a-Si has a direct band gap of about 1.75 eV. That means that it is transparent to the light below 1.75 eV, including light between 1.1 and 1.75 eV; and above 1.75 eV, it is far more strongly absorbing than is crystalline silicon. Only about 1 micron of a-Si is required to absorb most sunlight, in stark contrast to the 100 microns needed for crystalline silicon to do the same job. This property— strong light absorption—made a-Si a desirable thin-film material.

The electronic properties of a-Si are also far different from those of crystalline silicon. Whereas diffusion lengths in single-crystal silicon can be over 100 microns, in a-Si they are usually measured in tenths of a micron. Dangling bonds and other defects can be minimal in crystalline silicon; in a-Si, they can be overwhelmingly numerous. In fact, the poor electronic properties of pure a-Si led early investigators to assume mistakenly that it would never be valuable for electronics. They considered the use of a-Si for electronic devices tantamount to using sandstone to build a dam—hopelessly inadequate.

However, in 1969 it was found that mixing hydrogen with silicon during the deposition of a-Si films led to much better electronic properties. A group at the University of Dundee, in Great Britain, developed a new technique called glow discharge that incorporated hydrogen with silicon. Glow discharge is a method in which a gas called silane (SiH_4) is mixed with hydrogen and both are broken down within a chamber in such a way as to deposit a mixture of silicon and hydrogen on a cell substrate. The amorphous silicon that forms is actually an alloy of silicon and hydrogen. The hydrogens attach themselves to defects in the silicon such as dangling bonds, passivating them.

The presence of hydrogen is critical to making any sort of successful PV devices with a-Si. For instance, before hydrogen was added, the presence of defects overwhelmed any attempt to dope a-Si either n- or p-type. Dopants were simply absorbed by the material in such a way that their electronic effects were neutralized by the dangling bonds. No excess carriers—holes or electrons—became available. But after hydrogen incorporation, the addition of traditional silicon dopants such as boron (p-type) and phosphorus (n-type) allowed the a-Si to be made n- and p-type. This was a critical advance without which a-Si would not have been usable as a thin-film material.

Once a-Si was improved by hydrogenation, interest grew in using it in solar cells. Several properties made it attractive: a band gap of 1.75 eV that is nearly optimal in terms of its match to the solar spectrum, a direct band gap suitable for thin-film cells, the use of silicon, a familiar and plentiful material, and a method (glow discharge) that could be adapted to make very uniform, large-area modules. Within a very short time (1972), RCA laboratories began work on a-Si as a thin-film PV material. They made the first a-Si solar cells in 1975. Between 1975 and 1980, more than 30 other groups in the US and Japan began working on a-Si, making it a leading thin-film material at the same time that serious difficulties were being encountered with copper sulfide.

Making a-Si Devices

The glow-discharge method (Figure 25) of making a-Si depends on the use of a plasma to break-down silane gas for deposition within a vacuum chamber. A plasma is an oscillating gas made of energetic electrons. The electron gas is capable of tearing molecules like silane apart. Once broken down in the plasma, the silane deposits both silicon and hydrogen on a substrate held at about 250° C. This temperature is chosen because at higher temperatures, the a-Si loses hydrogen and becomes polycrystalline. At lower temperatures, a-Si loses its proper electronic properties. The glow-discharge system pictured in Figure 25 has three sequential chambers in which different portions of the a-Si cell are deposited. The main feedstock gases are silane (SiH_4), phosphine (PH_3; n-type dopant), and diborane (B_2H_6; p-type dopant).

Properly grown, the a-Si material forms tall columnar structures similar to crystal grains. The columns stretch from the bottom of the layer to the top. This orientation facilitates charge movement during operation, since the free carriers being separated by the junction travel perpendicular to the cell surface, avoiding the boundaries of the columnar structures.

Substrates can vary, depending on the intended structure of the cells. The most frequent substrate is glass. (It is also called a superstrate since it ends up on top, with the light shining through it to get to the cell.) The first layer deposited is a transparent oxide—usually tin oxide. Tin oxide is highly transparent *and* highly conductive, so it can

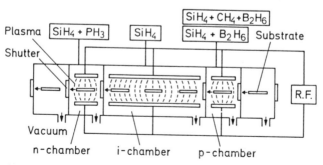

Figure 25. A three-chamber, radio frequency glow-discharge system for fabricating amorphous silicon.

act as a top electrode, covering the top surface of the cell without obscuring it. This is the same transparent contact that was developed for the copper sulfide devices.

Like all other PV devices, the key aspect of an a-Si device is its internal electric field. The PV structure that we are familiar with has n- and p-type layers that induce an electric field. Such a cell is called an n–p cell. But a-Si devices have evolved toward a relatively unusual structure called a p–i–n cell (Figure 25). The i-layer is an *intrinsic* layer, so named because it is not purposely doped either n- or p-type. It has a minimum of free carriers of either type.

The p–i–n structure evolved because the simple n–p structure did not work: n–p cell efficiencies were very low (under 5%). Both n- and p-type a-Si are heavily laced with defects that are introduced during doping. Only about 1% of the dopant atoms of boron or phosphorus act as dopant; the rest act as unwanted impurities, creating defects. The doped layers have poorer electrical properties than those of intrinsic material. Making a simple n–p a-Si cell from doped layers always resulted in a low-efficiency device with many defects in the electric field region.

To avoid this, the a-Si cell structure evolved toward one with three distinct a-Si layers. At first, it was an n–i–p design rather than a p–i–n design, in the sense that the top layer (and the first deposited after the tin oxide) was n-type rather than p-type. The n-type a-Si was very highly conductive and very thin (0.005 micron). It was made so thin in order to allow almost all of the sunlight to pass through it as if it were not there. Subsequently, a thicker, light-absorbing layer of un-doped a-Si was deposited. This i-layer was from 0.4 to 2 microns thick—enough to absorb most of the sunlight. Next, another very thin (0.005 micron) layer of a-Si was deposited. This one was p-type, the opposite of the first thin layer.

The heart of thin-film PV is the electric field that separates the oppositely charged electrons and holes. But an n–i–p design can also accomplish this separation. In a well-made a-Si device, an electric field was induced across the i-layer by the n- and p-type layers sandwiched around it.

The undoped middle layer must be nearly free of electrons or holes for a good built-in field to be induced. Recall that when n- and p-type materials are next to each other, they exchange excess majority carriers, inducing an electric field. When an undoped region intervenes between them, the same exchange can take place. But instead of the electrons from the n-type region entering the p-type region, the intrinsic region is depleted of its electrons and holes. Even in the purest intrinsic a-Si there are some free electrons and holes. The sources of these are ionized impurities, which donate free carriers.

Let's consider what happens when an intrinsic region is adjacent to an n-type layer. The holes in the intrinsic region absorb some of the excess free electrons from the n-region. A thin region of the n-side becomes positively charged, just as it would in an n–p structure.

Now let's transfer our attention to the other side of the i-region, next to the p-region. On this side of the i-region, the few free electrons in the i-region slip over into the p-type region, which is teeming with holes. A negative charge density is induced in the p-region, again, as it is in the p-side of an n–p junction.

The net result is that the i-region loses almost all of its free carriers while the n- and p-type regions around it become charged just as they do in an n–p structure. In physical terms, an electric field exists across the i-region, induced by the charges built up in the n- and p-layers sandwiching it. A light-generated free electron within the i-region would be pushed away from the p-side toward the n-side; and the simultaneously generated hole would be moved from the n-side to the p-side.

Since the a-Si is quite a strong absorber of light, even a single micron of intrinsic region is enough to absorb almost all the light. Because the entire i-region acts like the field region of an n–p device, nearly every solar photon can contribute to the electric current. This makes a-Si devices capable of nearly perfect current generation. To complete the cell, a metal such as aluminum or silver is deposited on the back of the cell, in direct contact to the p-type layer of a-Si.

Along with homojunctions and heterojunctions, the n–i–p design has become a standard PV device structure. Besides a-Si, it has also been adopted in other PV technologies, especially in cadmium telluride.

Improved a-Si Designs

In most a-Si cells now being made, the n–i–p structure has been reversed, becoming a p–i–n structure. The p-layer is deposited first on the tin oxide, then the i-layer, and the n-layer, last. This reversal evolved because it was found that a stronger electric field resulted between a p-type top layer and the i-region than between the n- and i-regions. Even in a-Si devices, most light is absorbed near the top, so it is best to have the strongest field in that region to facilitate electron–hole separation. Therefore, the p–i–n structure was adopted.

A stronger field exists between the p- and i-layer than between the n- and i-layer because undoped a-Si actually tends to be very slightly n-type due to the presence of unwanted impurities. A stronger field is induced between the p- and slightly n-type layer than between a doped n-layer and the undoped, but slightly n-type, "intrinsic" region.

The p–i–n structure introduced an added complication. To make it, p-type a-Si was deposited on n-type tin oxide. This would normally result in an electric field at the p–n interface of these two materials. The induced field would be in the wrong direction, opposite to the field across the i-type a-Si. The field would act as a barrier to the movement of current in the cell.

Nonetheless, the problem can be avoided by making both the a-Si and tin oxide highly conductive. Under such circumstances, the field that is induced becomes very narrow—less than 0.001 micron. This is a narrow enough region that electrons passing through it are not influenced by the field. They do not feel it, much as we do not feel the roughness of a "smooth" surface such as glass—despite the fact that at the submicron level, glass is not smooth at all.

When a field is induced between two highly conductive layers of opposite types, the carriers that are exchanged during field formation do not go very far into the opposite materials. There are so many locations near the interface that the exchanged free carriers cannot fill them all. The induced field region is very narrow and will not interrupt the movement of electrons.

This kind of interface, where an opposing built-in field does not actually impede solar cell performance, is called a tunnel junction because the electrons are said to tunnel across it as if it were not there. Good tunnel junctions are in a sense the opposite of good built-in fields: we want them to be ineffective in separating electric charges. Being able to make them is an important strategy for making many other advanced structures such as multiple cells stacked on top of each other (multijunctions). In this case, the tunnel junction allows us to use the p–i–n design because it allows the deposition of p-type a-Si on highly n-type tin oxide without there being a barrier opposing the flow of electrons into the cell.

Another phenomenon also favors tunnel junctions. Highly doped p-type a-Si has many defects. The narrow barrier between the n-type tin oxide and the p-type a-Si has many defects, which also act to allow the free flow of charges.

The Japanese Presence

One of the most important forces behind the progress of a-Si has been the strong participation of Japanese researchers. A combination of circumstances led to their entry into the field. Even though a few Japanese labs had investigated a-Si as an electronic material in the 1960s in parallel to the earliest work in the US and Great Britain, a-Si did not attract much attention as a solar cell material. When the Japanese Sunshine Project (their equivalent to our Department of Energy's National PV Program) started, it concentrated almost exclusively on crystalline silicon materials. But in the late 1970s, events in the US caused Japanese planners to seriously consider a-Si as an alternative for PV.

A US company, Energy Conversion Devices (ECD) of Troy, Michigan, was among those pioneering a-Si materials for electronics and for PV uses. In the late 1970s, Atlantic Richfield purchased a small PV company which they renamed ARCO Solar. It was later to become a world leader in PV. To get started in a-Si, ARCO Solar formed a joint venture with ECD, which included a $20 million investment from ARCO Solar to ECD. The interest of ARCO Solar, ECD, and other US companies—RCA and then Solarex—as well as the growing participation of SERI in funding a-Si R&D, convinced the Japanese to emphasize a-Si as their leading candidate thin film.

The Japanese quickly stamped the effort with their own characteristic ability to market new high-tech products. Japanese companies pioneered the use of small a-Si cells in pocket calculators. The fit was perfect. Light-powered calculators soon took a substantial part of the market from battery-powered calculators, providing a small but very high-value market for early a-Si cells. The consumer PV market was born.

The use of a-Si cells to power consumer products worked because the power output of these cells very well matched the microwatt power requirements of calculators. Meanwhile, the fact that the cells could be rather inexpensive compared to crystalline silicon cells allowed a-Si to seize this niche market. Having such markets has become important. It has provided a-Si with near-term markets upon which to build toward the ultimate market—large-scale power use. Subsequent to the calculator, other Japanese and US manufacturers have introduced other products—PV-powered watches and radios, outdoor lights for walks or patios, lights for illuminating house numbers, and many more. Each has provided some much needed revenue to companies developing a-Si.

The US Response

The Japanese emphasis on a-Si led to a US reaction. The SERI program to support a-Si expanded rapidly in the early 1980s. Ironically, our position was perceived to be one of catching up in a technology that we had pioneered. In 1983, an a-Si project office was formed at SERI. The first of several a-Si initiatives was begun by the federal government through SERI. A special contractual arrangement called a government–industry partnership was adopted for this initiative. The partnership was based on resource sharing. The DOE, through SERI, provided each company with 70% of contract funds; the company provided the remaining 30%. The rationale for these partnerships—four of which began in 1983—was that:

☐ PV R&D was worthwhile from a national perspective.
☐ The R&D needed to accelerate the a-Si technology was very high risk. Without government stimulation, it would not have been conducted at a high enough funding level to be effective. This was especially true because of the financial situation of PV: it was not (and still is not) a money-making endeavor capable of sustaining its own R&D.
☐ Only companies willing to risk their own capital by sharing a portion of the contract's cost were considered sincere enough in their ultimate interests to be worthy of government funding.
☐ By funding *several* companies, competition was increased and research progress accelerated.

This partnership approach was not pioneered by the a-Si project, but as a strategy it was raised to a new level of funding and research intensity. Government–industry partnerships have since become the cornerstone of most SERI programs. Wherever possible, they have been reproduced in other PV technologies. In terms of meeting stated goals, they have been extremely effective. The a-Si research project at SERI recently (May 1989) received a prestigious award in technology transfer from the Federal Laboratory Consortium to recognize the project's achievements. The Federal Laboratory Consortium is made up of more than 500 government laboratories from 14 federal agencies. With the assistance of the a-Si research project, the US now dominates a-Si technology.

Module Development

One of the stated purposes of the first SERI initiative in a-Si was to accelerate the development of a-Si modules. The reasons were twofold:

☐ The Japanese were ahead of the US in pilot production of a-Si devices. Their lead was considered dangerous in terms of their perceived effectiveness in introducing new a-Si products.

☐ Requirements for large-scale power meant that modules of reasonable sizes (1 to 10 ft^2) were needed.

Based on a national competition, SERI funded four groups at over $1 million annually: 3M, Solarex (a subsidiary of AMOCO that had bought the RCA group working in a-Si), Chronar (Princeton, NJ), and Spire Corporation (Bedford, Ma). By the end of the first initiative (1986), Solarex was able to demonstrate a world's record 8%-efficient a-Si module with an area of 1 ft^2. This was (at the time) both the largest and most efficient thin-film module ever made, surpassing all achievements in copper sulfide. At the same time, a-Si technology and research spread widely throughout the US industrial and university communities. Cells of over 10% efficiency were being made by a dozen groups, and dozens more were interested in contributing to a-Si R&D.

New Directions

The first a-Si initiative established important milestones. It also identified several new critical research directions. Two very significant and related facts became clear:

☐ The simplest, single-junction a-Si devices suffered from a serious tendency to lose efficiency as they aged outdoors.
☐ A new kind of device design—called a multijunction—could be both less unstable and more efficient than the simple single-junction design.

Although a-Si cells over 10% efficiency were being made by 1985, they were unstable when exposed to sunlight. Within a few months, such cells would lose from 20–50% of their initial efficiency. This instability was called the Stabler–Wronski effect after the RCA scientists who were its discoverers. By the end of the first SERI initiative, the Stabler–Wronski effect came to dominate thinking about a-Si. Unless it could be avoided or ameliorated, the technology was doomed to the same fate as copper sulfide—oblivion.

The second SERI/DOE initiative in a-Si emphasized solving the Stabler–Wronski effect and developing high-efficiency multijunctions. It was started in 1986. Another four contracts were awarded, this time each for $1.5 million of government funds annually. The participants were ARCO Solar, Energy Conversion Devices, Chronar, and Solarex. Each matched SERI funds dollar-for-dollar, so that total contracts were about $3 million annually for each company. In every case, contracts were made to develop multijunctions, and to ameliorate or avoid the Stabler–Wronski effect.

As a result of these contracts and corollary progress within the same companies, several achievements of import were soon accomplished:

☐ The highest efficiency a-Si single-junction cell (12%) was developed by Solarex.
☐ The highest efficiency a-Si multijunction cell (13.3%) was developed by ECD.

□ The highest efficiency a-Si module (1 ft^2; 9.4%) was developed by ARCO Solar.

□ The largest thin-film module (10 ft^2) was developed by Chronar.

□ Several new products (e.g., outdoor patio light, address lights) based on a-Si were developed by Chronar.

□ Chronar and others started the initial large-scale production of thin-film modules.

□ The design, encapsulation, and testing of thin-film modules began.

□ The development of automated methods began for moving large-area thin-film modules to allow for large-scale, low-cost production.

□ By a fortuitous accident, the initiative supported and accelerated the development of an important, competing thin-film technology, copper indium diselenide (CIS), through funding of a contract with ARCO Solar. The contract was for an advanced multijunction design in which CIS and a-Si were combined as a multijunction concept. The success of this contract helped push CIS into the mainstream of PV. (See Chapter 10.)

Multijunctions

As stated previously, multijunction a-Si devices have the potential to be more efficient and stable than single-junction a-Si devices. We have seen how multijunctions work. For a two-junction device, high-energy light is absorbed in the top cell. Low-energy light passes through the top cell and is absorbed in the bottom cell. This means that the solar photons are used more efficiently. The high-energy portion of the spectrum is being used in a high-band-gap (high voltage) cell; but low-energy light is not being sacrificed—it is absorbed and used in a low-band-gap cell.

In true multijunction cells, the top cell has a high-band-gap absorber material; the bottom cell, a low-band-gap absorber. These cells are designed to split the solar spectrum evenly between them. But the first a-Si multijunctions were made using two cells with the *same* band gaps because the sophisticated technology needed to make materials of different band gap was not yet available. To match currents between cells (a requirement for optimal efficiency), the top cell of a-Si was made with a very thin (under 0.4 micron) i-layer. The a-Si was thin enough that about half the light was not absorbed in it and went through to the bottom cell.

This peculiar design was adopted because it was observed by some of the earliest a-Si researchers that a-Si cells with thinner i-regions had improved stability. So they naturally tried to make thinner cells. But in the process, they lost efficiency because some sunlight was able to penetrate the thinner a-Si layer without being absorbed. The next step was obvious: make another a-Si cell of the same band gap under the first. In so doing, they found how to match currents between the cells and could make multijunctions that were just as efficient as single-junction cells, but more stable.

During the second SERI initiative, new a-Si materials were developed that could be used for more efficient multijunctions. Mixtures of a-Si and carbon could be made that had higher band gaps than a-Si alone. Other mixtures with germanium were made with lower band gaps than a-Si. True multijunctions were developed that were capable of improved efficiency in comparison to single junctions because they could split the solar spectrum into its high- and low-energy halves. By putting them together, groups like those at ECD were able to surpass the efficiency of the best single-junction cells.

Besides increased efficiency potential, the main thrust of developing multijunctions was to ameliorate the Stabler–Wronski effect. Researchers knew thinner cells were more stable. Soon they found out why. The Stabler–Wronski effect occurs because the electronic properties of a-Si are reduced somewhat when a-Si is exposed to sunlight. Light produces free carriers that are caught by defects in the i-region, reducing the effectiveness of the electric field across that region. Introduction of these barriers to the free flow of charge across the i-region is somewhat analogous to the introduction of hazards in a shipping channel. The ships that would otherwise sail right through the channel are delayed or even damaged. The result of the Stabler–Wronski effect is a loss of efficiency in a-Si devices. For early devices, the loss was devastating: about 20 to 50% of initial performance was lost within months of outdoor use.

Because the effect is to reduce the field strength across the i-region, anything that *increases* field strength can counteract and minimize it. The field across the i-region is induced by the sandwiching n- and p-layers. By making the i-region thinner, the field across it is increased proportionally. In a sense, the problem behind the Stabler–Wronski effect is unchanged. The same charging of defects occurs when light is absorbed in the region. But the relative magnitude of the defects (compared to the increased field strength) is smaller. Proportionally less degradation occurs.

One would like to make cells thinner and thinner to obviate the Stabler–Wronski effect completely. But as we know, using thinner layers in *single*-junction devices allows too much light to pass through without being absorbed. Current—and efficiency—shrinks precipitously.

Two approaches to efficient, stable cells were developed: putting mirrors on the back of single-junction cells and developing multijunctions. Mirrors, like light-trapping in crystalline silicon cells, can be very effective in increasing current by allowing light to pass more than once through the absorber region. This is still a promising approach, although materials limits have slowed it. For instance, the best mirror is silver, which is too expensive.

But multijunctions can be an even better approach, because they have a higher potential efficiency if fully developed. Thus, the existing effort to optimize a-Si for both efficiency and stability concentrates on multijunctions. The work to develop new a-Si alloys with carbon and germanium has consumed several years of funding and is now bearing fruit with the attainment of very high efficiencies (13.3% at ECD). Fortunately, multijunction a-Si devices are relatively easy to make because different layers can be made with the same glow-discharge process simply by introducing different feedstock gases into the chamber. These feedstock gases carry the alloying material—carbon or germanium—as well as the silane needed for the a-Si. In fact, the multiple layers of a complex a-Si multijunction can be made in almost the same time as the a-Si layers of a single junction because the total layer thickness is about the same for both. More steps are involved (which adds to the time) and yields are impacted, since the multiple layers require greater variation and more subtle control. The ability to make multijunctions relatively easily is a key advantage of a-Si and its alloys. None of the other thin films can claim the same flexibility.

Safety

Although silicon is itself a rather benign material, the feedstock gases used to make a-Si devices are not. Several of them—silane, diborane, phosphine, and germane—are among the most dangerous gases in use today. Silane (which is used in even greater quantity preparing silicon for crystalline silicon cells) is explosive and highly toxic. The others are highly toxic. The semiconductor industry uses all of these gases and has proven itself capable of handling them safely despite their toxicity. No worker deaths in PV have resulted from their use to this date. But it is fair to say that safety issues—not usually associated with PV by the general public—will have to be handled carefully as a-Si or other PV materials are scaled up in proportion to rising markets.

Impact of Instability

Stability may be the key issue determining the future of a-Si. Besides the thin cells now being developed to minimize instability, another factor affecting stability has been found: Light-degraded a-Si cells can be heated at under 100° C, and their original efficiencies can be restored. The same phenomenon affects modules outdoors. Operating temperatures in the range of 50° C are enough to cause a slow self-annealing of a-Si modules while they are outside. Especially during the hot summer months, the self-annealing can raise the efficiency of degraded a-Si cells toward their original values. This tends to put a limit on the degradation from the Stabler–Wronski effect.

Unfortunately, other instabilities have also affected the first generations of a-Si modules outdoors. Design of a-Si modules and their encapsulations are rather early in their development; and even crystalline silicon modules were unstable in the 1950s and 1960s when first put outdoors—because of engineering problems with early designs. The same debugging process of module designs and encapsulations will necessarily have to be done to optimize a-Si or any thin-film modules. But because degradation of a-Si modules outdoors has been large, no one can say with certainty that the intrinsic degradation of the Stabler–Wronski effect is under control. This uncertainty haunts present efforts.

Manufacturing Cost

Several manufacturers of a-Si panels have made projections of the cost of manufacturing their panels in large volume. These projections generally agree that four conditions are necessary to make low costs an achievable goal for a-Si:

1. Control of the Stabler–Wronski effect to a 10% loss
2. 10%-efficient modules (after degradation) of 0.5–1 m^2 area
3. Annual production of at least 10 megawatts
4. An automated production line

Assuming that these criteria will be met, the a-Si manufacturers also analyzed module costs, dividing them into the following categories: material, depreciation, labor, and indirect costs. One study by Dave Carlson, a Vice President at Solarex Corporation, made the following estimates of material costs (assuming a 90% module yield and 1-micron-thick a-Si layers):

Encapsulant	$11/$m^2$
Glass	$6/$m^2$
Germane gas	$5/$m^2$
Silane gas	$3/$m^2$
Other (tin oxide)	$6/$m^2$
Total	$31/$m^2$

Depreciation of equipment is another major cost. (Depreciation is another way of saying the cost of the capital equipment over the life of the manufacturing facility.) Solarex assumed, based on their experience with equipment for their existing a-Si lines, that they would need about $10 million for a 10-MW annual production capacity. For a ten-year plant life, their annual cost of equipment would come to be about $11/$m^2$.

Because the Solarex estimate assumed an automated plant in which glass is loaded at one end of the line and panels are fed off the other end, labor was minimal. Only 24 people were employed by the 10-MW plant, 8 per 8-hour shift. The labor cost was $4/$m^2$.

Indirect costs included electricity ($4/$m^2$), and other costs such as rent, waste disposal, maintenance, insurance, totalling $4/$m^2$; i.e., for total indirect costs of $8/$m^2$.

The total cost of these components is about $55/m^2. Using the same assumptions as previously for BOS cost, this translates to a cost of energy to the consumer of about 6.6 cents/kWh$_{DC}$ for 10%-efficient panels. This should not be seen as a final estimate of such costs. Future, stable efficiencies could be higher than 10% for a-Si modules, and costs could also come down as manufacturing capacity grows.

Other Issues

If the instability of a-Si devices can be controlled and losses are only about 10% over the life of a device, the Stabler–Wronski effect can be tolerated. Any loss is a problem, but with sufficient progress in efficiency, a 10% loss can be overcome. The onus would be on achieving very high efficiencies. At this point, good efficiencies have been achieved with three-junction cells made by ECD (up to 13.3%). More progress is needed, perhaps efficiencies up to 18% for cells, to achieve truly low-cost electricity. In addition, these best results will have to be translated into module efficiencies, an accomplishment that has yet to be realized. The best a-Si modules are now in the range of about 8% (after 10% degradation), and typical modules are even lower—5% or so. The technology for making large-area multijunction modules has not yet been put in place. The demonstration of high-efficiency a-Si modules awaits more advanced production techniques.

Competition

Another factor affecting the future of a-Si is competition from other PV technologies. While a-Si has been progressing at a reasonable rate, other thin-film technologies based on CIS (Chapter 10) and cadmium telluride (Chapter 11) are developing as well. Despite substantial funding over the last six years, a-Si is no longer considered the dominant thin film. The other thin films also have their adherents. The fact that their impressive progress has come despite only minor funding by government and industry means that they may be more suitable PV materials. Technical problems with these materials are not as intractable as those encountered in a-Si. Increased emphasis on them should result in rapid progress. Some of the problems of a-Si (instability, efficiencies below 10% for modules) would not seem so glaring were it not for the fact that other thin films are progressing beyond such issues.

10 ☼ CIS

A single-crystal copper indium diselenide ($CuInSe_2$ or CIS) cell of 12% efficiency was made at Bell Labs in 1975, spurring interest in this rather exotic compound semiconductor. Since then, a small effort has been directed at developing it. Several groups—University of Maine, Boeing Aerospace Corporation, Solar Energy Research Institute, Institute of Energy Conversion (University of Delaware), ARCO Solar, and International Solar Electric Technology—have been enthusiastic about the potential of CIS. Others have been preoccupied with silicon—crystalline or amorphous—and felt that CIS was a minor material of little value to PV.

But in the last two years—powered by a strong research program at ARCO Solar—CIS has emerged from the shadow of other PV options and has become perhaps the leading PV candidate for future large-scale applications. Through CIS, the power of the thin-film approach is finally becoming real.

Beginnings

In the mid-1970s, a small group at Bell Laboratories in New Jersey began working on several unusual compound semiconductors for advanced optoelectronic applications. Among the materials studied were $CuInSe_2$ and some related compounds called chalcopyrites ($CuGaSe_2$, $CuInS_2$, and others). Responding to growing interest in PV following the 1973 oil crisis, Bell Labs began experiments using the chalcopyrite semiconductors to fashion single-crystal solar cells. The single crystals were grown as boules in much the same way as silicon, e.g., by the Czochralski process. Then the boules were sawn into wafers, and cells were fabricated on the exposed surfaces. Almost immediately, Bell Labs had success with one of the materials—Cu-InSe$_2$—reaching 12% efficiency.

CIS—as this complex compound semiconductor is sometimes called—is a low-band-gap material (about 1 eV) and is the most light-absorbing semiconductor known. Half a micron of CIS is sufficient to absorb 90% of solar photons. This quality makes CIS a very suitable material for thin-film cells.

But low-cost CIS cannot be attained with single crystals grown by the Czochralski process. In order to make low-cost thin-film cells a rapid deposition process onto inexpensive substrates is required. Bell Labs' single-crystal results attracted interest in CIS, and other groups tried to make CIS cells on metal-coated ceramic or glass. But the use of non-single-crystal substrates naturally resulted in polycrystalline CIS and much poorer efficiencies.

Still, progress was rapid in making polycrystalline CIS cells. The University of Maine made a polycrystalline CIS cell of 6.6% efficiency in 1976. Though only half the efficiency of the original Bell Labs' single-crystal cell, this was an impressive result for an approach with low-cost potential.

The CIS was deposited on a metal-coated glass substrate by a method called vacuum evaporation. In vacuum evaporation, materials are heated and vaporized in a vacuum. The vapors then move in a line-of-sight direction to the desired substrate, where they deposit in the form of a thin film. Sometimes the materials react on the surface of the substrate to form the desired material; other times, a material of the desired composition is vaporized itself and re-formed on the substrate. In the University of Maine experiments, the source materials were CuInSe$_2$ and Se. Both were heated to evaporation in high vacuum. Copper indium diselenide was transported to the substrate in the presence of added selenium, which helped compensate for some loss of selenium from the CuInSe$_2$ during vaporization. The process required very careful control to achieve CuInSe$_2$ with the proper ratio of copper, indium, and selenium across the entire substrate and throughout its depth. Achievement of the right ratio of copper to indium to selenium (about 1 to 1 to 2 as in the formula CuInSe$_2$) is called compositional, or stoichiometric control and remains a central issue of all CIS fabrication.

Boeing

Although Bell Labs and Maine got CIS going, the most important early work in CIS was done by a small group at Boeing Aerospace Corporation in Seattle. Funded since 1976 by federal agencies (Energy Research and Development Administration and then DOE and SERI until 1989), Boeing startled the PV community by achieving over 10% efficiency with CIS cells early in 1982. At the time, their achievement was the best efficiency result among the various thin films and the second thin-film cell (after copper sulfide) to exceed the significant 10% mark.

Boeing's team, led by Reid Mickelsen and Wen Chen, used a variation of the Maine approach called three-source evaporation to make CIS. The source materials were elemental copper, indium, and selenium. These were heated in a vacuum to evaporation and co-deposited on a molybdenum-coated ceramic substrate. By today's standards, the Boeing evaporation method seems complex and slow, requiring careful control and an hour to make a 3-micron layer of $CuInSe_2$. The molybdenum that formed the back contact layer beneath the CIS was sputtered on a ceramic substrate. Sputtering is a method in which high-energy particles, usually electrons, are used to "knock off" atoms from a target material in such a way that the sputtered material deposits itself on a desired substrate. Molybdenum has since become the conventional back contact for CIS devices.

Figure 26 shows the entire Boeing cell structure. (For more detail, see below.) The CIS layer was made p-type. On top of the CIS, Boeing evaporated n-CdS to make a junction with p-CIS. Like the n-CdS/p-Cu_2S, this was a heterojunction device. Aluminum grids were deposited on the CdS as top contacts. For very high efficiencies, antireflection coatings such as zinc sulfide or magnesium fluoride were deposited on the CdS to minimize reflections.

Boeing's work resulted in several early CIS patents. One of the most important was for a so-called two-layer CIS film, which Boeing considered crucial to making high efficiencies. Understanding the

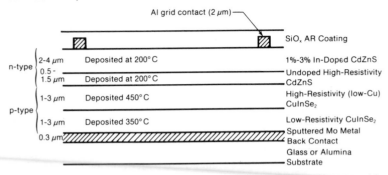

Figure 26. The Boeing CuInSe₂ cell structure with two-layer CuInSe₂. The compositions of the two layers were copper rich near the molybdenum; copper poor nearest the top of the CuInSe₂ layer.

Boeing two-layer approach requires more background on the nature of CIS:

CIS is a complex material. It can be doped p- or n-type by adding impurities, just like silicon. This is called extrinsic doping. But doping can be accomplished in CIS in another way: by altering the ratio of Cu to In to Se from a perfect 1:1:2 stoichiometry. This so-called intrinsic doping is done by adding to or taking away from the concentration of copper, indium, or selenium during deposition. Boeing, and most other groups since, have made CIS highly p-type by depositing relatively more copper than indium in the film. They have made CIS n-type by the reverse: more indium than copper. In most cases, the selenium content remains relatively steady, near 50% of the total number of atoms.

In terms of the lattice structure, this means that for n-type CIS material, some copper atoms are missing from the lattice, perhaps replaced by extra indium atoms. An indium atom located on a copper site in the CIS lattice tends to contribute electrons to the conduction band because indium has two more outer electrons than copper does. The extra electrons contributed by numerous indium atoms on copper sites make the material n-type. On the other hand, when *copper* is in excess and takes positions on the indium site, copper has too few electrons for its position in the lattice and tends to grab electrons, making the CIS p-type. These examples show how intrinsic doping works. It is also a brief glimpse of the complexities of a three-element compound semiconductor like CIS.

The Boeing patent on two-layer films concerned the deposition of CIS films where the first layer deposited was intentionally made p-type and the second layer—nearest the CdS interface—was either highly resistive or slightly n-type (see Figure 26). During the deposition of the p-type layer, more copper than indium was evaporated and temperatures were held at 350° C; for the second layer, the copper deposition was slowed, so that more indium than copper arrived on the substrate. Substrate temperature was raised to 450° C. Boeing attributed their high-efficiency results to this two-layer approach.

At first, one of the main benefits of the two-layer approach was that it avoided a known problem. In CIS films that were close to perfect stoichiometry (25% copper, 25% indium, and 50% selenium), copper nodules formed at the interface between CIS and CdS. This was reminiscent of one of the catastrophic degradation mechanisms observed in Cu_2S/CdS cells and was very discouraging. The effect in CIS cells worsened if there was an excess of copper in the CIS near the CdS interface. By depositing two-layer CIS, with a dearth of copper near the CdS (under 25%), copper nodule formation was suppressed. Boeing not only avoided this catastrophic problem but made efficient cells using the same strategy.

Subsequent investigations of two-layer films, especially those studies carried out at SERI by Rommel Noufi, strongly suggested that the copper- and indium-rich layers mixed almost completely during deposition. A CIS film that has been made through evaporation as two distinct layers becomes almost homogeneous. However, mixing is not perfect. At the rear of the cell, closest to the molybdenum, a compositionally graded CIS layer remains that is copper rich and strongly p-type. This highly conductive layer makes a good contact with the molybdenum. On the other hand, throughout most of the bulk of the CIS, the copper and indium *are* fully mixed, yielding a slightly p-type and very nearly perfect CIS layer in terms of atomic ratios (Cu = 1:In = 1:Se = 2). However, within a quarter of a micron of the CdS interface, the material remains slightly *indium* rich. Indium-rich CIS does not have any extra copper to form nodules.

Boeing observed that cells made without their two-layer approach exhibited very poor efficiencies compared with those made with it. The two-layer CIS structure improved Boeing cell performance for several reasons. We know that an electric field is induced by an n–p junction such as the one between n-CdS and p-CIS. The placement of a resistive (indium-rich) region in the CIS nearest the CdS produced an n–i–p-like structure in that region in much the same way that an intrinsic a-Si middle layer produces an n–i–p structure in a-Si cells. With a resistive layer next to the n-CdS, an electric field stretches across it to the more p-type Cu-rich CIS layer underneath. This allows for a wider electric field region. CIS can absorb 90% of the sunlight in less than half a micron of material. So almost every solar photon is absorbed in the enlarged field region, making nearly perfect currents possible.

The two-layer approach allowed Boeing to use copper-rich material near the back of the cell where the resulting p-type material made good contact with the molybdenum; it allowed them to have resistive, copper-poor material near the junction to provide a wide drift field while avoiding unwanted copper nodules; and it provided one other advantage: better grain growth. Material with more copper tends to form larger grains. Thus, by starting with a first layer (nearest the molybdenum) of copper-rich material, larger CIS grains were formed. Indium-rich layers naturally have smaller grains, but grown on top of copper-rich layers they tend to conform to the same growth pattern— larger grains. Thus, the two-layer approach provided a number of advantages that translated into higher efficiencies.

Current was not the only characteristic improved by the two-layer approach: voltage also improved. Single-layer films that were slightly copper rich tended to have poor voltages. It was not well-known at the time, but such films were made of a combination of $CuInSe_2$ and another material—copper selenide (Cu_2Se). Copper selenide has many of the bad attributes of its closely related cousin, copper sulfide (Cu_2S). Both tend to have copper-nodule formation associated with their use with CdS. The nodules formed in copper-rich $CuInSe_2$ were actually due to the presence of a small concentration of Cu_2Se. Thus, the junctions in copper-rich, single-layer $CuInSe_2$ films were actually CdS/Cu_2Se junctions, with very poor voltages. By reducing the copper content near the CdS and making true $CuInSe_2$, voltages were improved.

But the voltage improvement was related to another complex phenomenon that was also discovered by Boeing: their cells required a short heat treatment in oxygen or air (20 min at 200° C) to reach their full efficiencies. The heat treatment was not a minor processing step.

Two-layer cells that were only 5% efficient before heating in air could be 10% efficient afterwards.

The Boeing anneal has been studied extensively since its discovery in about 1979. Much of the recent work has been by a team at SERI led by Rommel Noufi. They found that oxidizing treatments generally improve CIS cell efficiencies; and chemical reductions decrease efficiencies. The treatments are reversible, in the sense that the original cell efficiency can be reproduced by reversing the chemical process sequence. The effect of chemical treatments on CIS is complex, but some researchers believe that oxygen passivates defects within the two-layer CIS, especially those in the electric field region. The defects are those introduced by an excess of indium in the indium-rich CIS region. They believe that heat treating two-layer cells improves voltages because defect concentrations are reduced substantially in the crucial electric field region.

Stability

The most important early discovery associated with two-layer CIS cells was that they seemed stable. (As stated above, those that were copper-rich at the CdS interface were unstable due to the formation of copper nodules caused by the inadvertent presence of copper selenide.) Early two-layer Boeing cells could be stored or exposed to light for very long periods without any degradation. Tests as long as 8000 hours under illumination were done as early as 1982 (Figure 27). The stability of CIS was a unique characteristic and the major reason that the thin-film program at SERI abandoned all work in copper sulfide early in the 1980s in favor of CIS. Subsequently, SERI has been a major supporter of CIS R&D.

Based on their rapid progress, Boeing began a joint venture to commercialize CIS for terrestrial PV in 1983 with an oil service company, Reading and Bates. Sovolco, as the venture was called, was short-lived. Financial difficulties induced by plunging oil prices forced Reading and Bates to abandon the joint venture. Boeing's major effort to commercialize CIS was at a standstill. During the short-lived program, however, they made impressive progress in scaling up their evaporation process to make the first large-area CIS submodules. These were still rather small by today's standards (100 cm^2), but effi-

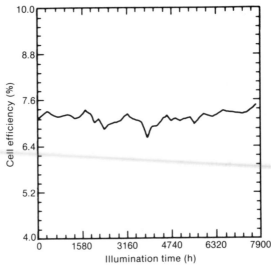

Figure 27. Boeing helped establish the intrinsic stability of $CuInSe_2$ devices with tests of unencapsulated cells as early as 1982. These tests showed no degradation for up to 8000 hours under simulated sunlight indoors. (Small variations in the data are due to measurement uncertainty.)

ciencies of 6% were demonstrated. At the time, these were among the most efficient thin-film submodules of their size.

But the failure of Sovolco dealt a severe blow to Boeing's research team. It blocked their route to commercialization of CIS as a terrestrial technology. Subsequently, they have survived as a research group only by developing CIS for use on satellites—the traditional use for PV, but not the original intentions of their group. Indeed, the collapse of Sovolco would have cost the CIS technology several years of delay except for the fact that in the meantime, the world's largest PV company, ARCO Solar, had begun serious work on CIS. Although work continued at Boeing and at SERI and the Institute of Energy Conversion (part of the University of Delaware), the ARCO Solar effort became the only clear corporate path to CIS commercialization.

ARCO Solar

ARCO Solar's interest in CIS dates back to the early 1980s. Along with their interest in a-Si, they started a program in CIS and supported it with their own funds. Their decision to investigate CIS was based on the efficiency and stability achievements demonstrated by Boeing. Their research team was led by Vijay Kapur.

At first, ARCO Solar contented itself with developing three-source evaporation in parallel to the work at Boeing. However, by 1983 a new, potentially lower cost and less complex process called selenization was developed by ARCO Solar. Selenization has since become the most important approach to fabricating high-quality CIS.

Selenization requires two steps and is sometimes called the two-stage process. In the first step, copper and indium layers are deposited on molybdenum-coated glass. In the second, the copper–indium layers are exposed to a selenium-bearing gas such as hydrogen selenide mixed with argon. The hydrogen selenide breaks down and leaves selenium, which reacts and mixes with the copper and indium in such a way as to produce very high-quality $CuInSe_2$.

Selenization requires a means of depositing copper and indium in a very nearly perfect ratio across a large-area substrate. Several methods have been tried. A favorite is sputtering. Sputtering is a method that is somewhat similar to glow discharge, the method used to make a-Si. In this case, however, the electron plasma touches a source material such as copper or indium. Unlike glow discharge, in which silane is introduced, no gas feedstock is introduced in sputtering unless a reactive species such as hydrogen selenide is needed. The electrons in the energetic plasma sputter atoms off the source material from its surface, and these atoms deposit on a desired substrate. Source materials of both copper and indium have been successfully developed for the first step of the selenization process.

Sputtering is an established process for very high-throughput manufacturing by the coating industry. It is used by this substantial industry to make thin metal layers on large, glass sheets. These are then used in office buildings to reflect sunlight and to keep the buildings cool. Companies expert at large-area sputtering are able to coat moving sheets of glass that are as wide as 13 feet. They use specially designed sputtering targets called magnetrons. Magnetrons are large, flat expanses of metal. When exposed to a controlled plasma, these metal plates produce very even beams of sputtered atoms capable of coating large areas uniformly and continuously. About 30% of the metal in the targets is used during deposition, but the rest can be removed and reused.

The development of sputtering and selenization for making CIS established the first viable method suitable for making low-cost CIS modules. ARCO Solar's development of selenization was an important step in the evolution of the CIS technology.

In 1984, the head of ARCO Solar's CIS R&D left the company and started one of his own: International Solar Electric Technology (ISET). They have subsequently become an important part of the CIS industry—the only other company (besides Boeing) with an established position in CIS.

Progress in Cell Efficiency

During the mid-1980s, progress in CIS cell efficiencies came almost exclusively from ARCO Solar and Boeing. ARCO Solar reportedly reached 12.5% by about 1984, demonstrating the highest efficiency thin-film cell of its time. The advance was made as a result of ARCO Solar's development of a new structure for the CIS cell (Figure 28).

The Boeing structure (Figure 26), which ARCO Solar improved, had an n-CdS top layer (also called a window layer) of about 3 microns

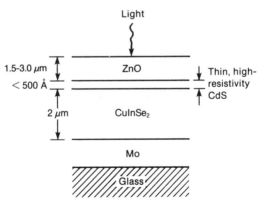

Figure 28. ARCO Solar developed a new design for the CdS/CuInSe$_2$ cell in which thin CdS (less than 0.05 micron) replaced thick CdS (3 microns). The thinner CdS allowed most of the ultraviolet light to reach the CIS and contribute to the electric current.

in thickness. But CdS has a band gap of 2.4 eV, which makes it a good but not exceptional top (window) layer from the standpoint of transparency. The CdS blocks photons having more energy than 2.4 eV, which is about 15% of the solar photons CIS cells could respond to. In the Boeing structure (thick CdS), these energetic photons were absorbed near the top of the CdS layer. The CdS was about 3 microns thick, so the electron–hole pairs thus generated were too far from the CdS/CIS electric field. As in other non-single-crystal materials, the diffusion length in CdS is very small. Electrons and holes generated by light at the top of thick CdS had almost no chance of contributing to the current.

ARCO Solar changed the structure to avoid this loss. Their advance was based on making the CdS thin enough—less than 0.03 micron (300 angstroms)—so that it could not absorb much sunlight. Instead of high-energy photons being absorbed 3 microns from the electric field, they went right through the thin CdS and were absorbed (along with the rest of the spectrum) at the top of the CIS, within the electric field at the CdS/CIS interface.

To make thin CdS, ARCO Solar developed a new method of depositing the CdS called dip-coating. Based on some work done in India by a group led by K. L. Chopra, dip-coating of CdS is based on precipitating CdS from a solution containing Cd and S molecules. It is capable of very good coverage despite layers that are so thin (100–300 angstroms) that even electron microscopes are incapable of delineating them. Like the ARCO Solar design, dip-coating of thin CdS has become the industry standard.

The ARCO design increased current by 10–15% over the Boeing design. Because efficiency is a product of current and voltage (and fill factor), higher currents increased efficiency by the same percentage. ARCO Solar completed its new design with a highly transparent, conducting layer of zinc oxide (ZnO) on top of the thin CdS. The zinc oxide was needed to carry the electric current laterally, across the top of the cell. Without it, the current flowing sideways through the thin, resistive CdS would have caused unacceptable resistive losses. Several methods have been adopted for making ZnO: sputtering from a ZnO target and a process in which vapors containing zinc and oxygen are mixed just above the substrate. To date, neither of these approaches is deemed perfectly satisfactory from the standpoint of cost.

Large Areas

Despite cells that were often the most efficient of any thin-film technology throughout the 1980s, CIS did not attract much new federal or corporate attention. CIS received only about $1–$2 million annually from SERI/DOE at a time when the silicon technologies (both crystalline and amorphous) received ten times more support and dominated federal and corporate efforts here and abroad. At first, skeptics said that CIS could not be made in a cheap enough way to be a low-cost thin film. They decried the Boeing method as being far too expensive and were unaware of the ARCO Solar selenization method, whose existence was a well-guarded secret throughout the 1980s (US Patent 4,798,660, January 17, 1989). Skeptics also said that CIS could not be made in large enough areas to be relevant for PV production. Until CIS modules were made, DOE/SERI and the fragile PV industry—already heavily invested in other technologies—refused to switch significant resources into CIS.

Boeing, during its Sovolco venture, was the first to make efficient 100-cm^2 CIS submodules with interconnected cells. They made 6.6% efficiency on these by 1985 and then 9.6% in December 1986. At the time, the 9.6% represented the highest efficiency for any thin film of that size. It was one of the key advances in CIS technology.

Although the Boeing result rekindled some interest in CIS, very significant attention was attracted when ARCO Solar succeeded in making breakthrough efficiencies on modules of one square foot size and larger in 1988 and 1989.

ARCO Solar developed CIS with internal funds through 1985. In 1986, they were among those awarded major research contracts from SERI through its second a-Si initiative (Chapter 9). They received $1.5 million per year, matched by another $1.5 million of their own money. The second a-Si initiative focused on a-Si multijunctions to improve stability and efficiency. ARCO Solar proposed a unique multijunction structure based on an a-Si cell placed on top of a CIS bottom cell. From a theoretical standpoint, the two cell materials are complementary in terms of their use of the solar spectrum. Amorphous silicon has a high band gap (1.75 eV) and CIS has a very low band gap (1.0 eV). Amorphous silicon absorbs almost half of the solar spectrum, and CIS absorbs the other half. They are almost a perfect combination for an efficient, two-junction cell. SERI accepted ARCO Solar's a-Si/CIS approach in the a-Si competition, and this provided a substantial increase in the funding of R&D in CIS at ARCO Solar.

Subsequent progress in CIS at ARCO Solar was rapid and impressive. During the course of the contract, ARCO Solar chose to reorient their SERI contract to concentrate their efforts on CIS. Their corporate reorientation to CIS is an example of a more general phenomenon that their success in CIS instigated: a general shift of the thin-film PV community away from a-Si and toward CIS.

Record-Shattering Results

In June 1988, ARCO Solar created a new standard for thin films by fabricating a world-record 11.1%-efficient square-foot CIS module. This was the first thin-film module to exceed the 10% barrier, which is two-thirds of the efficiency needed to make the DOE goal of 15%. Figure 29 shows the rapid progress of CIS modules in relation to other thin films. This rapid progress to record efficiencies is the force behind the recent spread of the CIS technology to companies such as Chronar

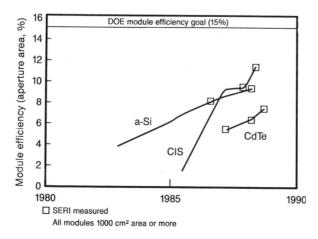

Figure 29. Copper indium diselenide modules (about 1000 cm²) fabricated by ARCO Solar have made recent progress toward world-record efficiencies among the thin films.

Corporation and Solarex, which have traditionally been identified with silicon. Prior to the ARCO Solar result, thin films had aspired to 10% efficiency as a distant goal. The ARCO Solar result represented the first real indication that thin films could do the job that they had been developed to do.

In addition to their high efficiencies, the ARCO Solar CIS modules have proved to be stable. ARCO provided SERI with two first-generation CIS panels in the summer of 1988. SERI tested them for more than half a year *with no degradation* (Figure 30). This is the first independent corroboration of the stability of *any* thin-film modules. Not only have CIS modules established themselves as the most efficient thin-film modules, they have been the only ones able to maintain their efficiencies outdoors. Global leadership in efficiency and stability has made CIS today's leading thin-film technology.

Subsequent to their 1-ft^2 module, ARCO Solar began to scale up to even larger areas. Six months after they began to make them, in January 1989, they produced a 9.1%-efficient 4-ft^2 CIS panel (four times the size of their earlier module). This was far and away the most efficient thin-film panel of its size. It produced 35.8 W, a power output approaching that of conventional crystalline silicon modules of the same size. No other low-cost thin film comes close.

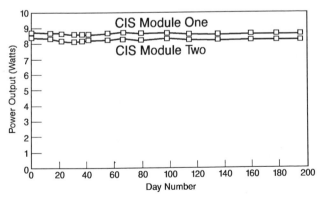

Figure 30. In 1989, two ARCO Solar CIS modules were tested outdoors at SERI for over a half a year without degradation. This was the first independent verification of the stability of any thin-film modules.

New Cell Efficiency Records

The DOE goal for thin-film modules is 15%. Prior to the ARCO Solar results in CIS, most in the PV community were skeptical of any thin film ever achieving the DOE goal. The record-breaking ARCO Solar modules were made using production techniques similar to the ones they used to make their 12.5% cell. The ratio of the efficiency of their 11.1% module to their 12.5% cell is 89%. This provides a measure that can be used to predict their future achievements in *module* efficiencies based on their progress with small cells. Assuming the same ratio (cell efficiency to module efficiency of 89%), a CIS cell of 17% efficiency would be needed to make the DOE goal of a 15% module.

Parallel to their improved module results, ARCO Solar was able to make substantial progress in the efficiency of their CIS cells by incorporating small amounts of the element gallium in place of the same amounts of indium. Since the mid-1980s, ARCO Solar, ISET, and Boeing had been investigating the use of gallium as a substitute for equivalent amounts of indium in CIS. This substitution produces an alloy of CIS with gallium, i.e., $Cu(In,Ga)Se_2$, sometimes called CIGS. Substituting gallium for indium in CIS raises the material's band gap, which moves the band gap toward more optimal values in terms of theoretical efficiency. We know that raising the band gap raises voltage but also decreases current. Between 1.0 eV (CIS) and 1.5 eV, the trade-off is favorable: efficiency increases with band gap because voltage increases faster than current falls.

Both Boeing and ARCO Solar found that adding small amounts of gallium for indium (up to 25% gallium in relation to indium) increased voltage even more than would be expected from increasing the band gap. Gallium seemed to act the way oxygen does, reducing the effect of defects in the CIS within the electric field region, allowing for higher voltages. By using gallium in its cells, Boeing made 12.9% efficiency in the spring of 1988. In the summer of 1988, ARCO Solar was able to achieve a record-breaking *14.1%* efficiency using gallium—well above any comparable value among the low-cost thin films.

The achievement of 14.1% efficiency raised general expectations about the potential of CIS to reach the efficiency needed to make 15% *modules*. Of course, modules are not cells: they are far bigger and much more challenging. But based on an 89% ratio of small-area cell efficiency to module efficiency, a 14.1% cell efficiency should lead to 12.5% modules when integrated into the ARCO Solar production process for larger areas. ARCO Solar has published a technical paper entitled "CIS: Towards 20% Efficiency," in which they give details of their expectations of reaching this exalted efficiency for single-junction CIS cells. At the 1989 meeting of SERI's CIS subcontractors, several presented arguments supporting the eventual achievement of 18- to 22%-efficient CIS cells. The potential for achieving the DOE goal of 15% efficiency for a low-cost module seems clear.

Issues

The CIS technology appears to be very promising, but it is not without some limitations. In terms of the classic measures of success—efficiency, cost, and stability—CIS is very strong, with clear leadership in efficiency and stability. But production cost remains an issue for three reasons: (1) a CIS deposition method based on sputtering requires high initial costs, (2) selenization is a slow process, and (3) one element needed to make CIS—indium—is expensive and produced in limited quantity.

Sputtering is a vacuum technique that requires costly equipment for very large areas. But it is capable of very high rates, somewhat ameliorating this problem. Sputtering rates of 1 micron/minute have been achieved for metals; such rates are expected in the deposition of copper and indium in CIS production. Since the metal layers in CIS are less than 1 micron thick before selenization, they can be sputtered in under a minute. On the other hand, selenization is a slow process, taking more than an hour. It is not a complex process, however, and is comparable to a heat treatment. Very large numbers of modules can be selenized at one time.

CIS manufacture has not been demonstrated or analyzed, so low cost remains somewhat uncertain. Most experts believe that manufacturing CIS, given existing techniques, is quite comparable to making a-Si; if the latter can be made inexpensively, CIS can as well. However, several companies are investigating alternatives to sputtering to reduce manufacturing cost. The methods being studied include electrodeposition of the metals from solution and spraying of a metals-containing solution. A small company, International Solar Electric Technology (ISET), in Inglewood, California, has been a leader in these investigations. The company was formed in 1984 (with the help of funding from SERI/DOE) by two key researchers from important PV companies, Vijay Kapur (from ARCO Solar) and Bulent Basol (from Monosolar; see next chapter). However, to this date no new process has yet replaced sputtering as the conventional method of depositing copper and indium prior to selenization.

Indium

Indium is the most expensive semiconductor material in CIS modules. It costs about $300/kg in large amounts of reasonable purity (0.9999 pure). CIS layer thickness in ARCO Solar cells is 2 microns, and there is 4 g of indium in a square meter of CIS that is 2 microns thick. Indium utilization in sputtering is about 30%, but some systems are 90% efficient, and unutilized material can be recycled. Using existing systems, the indium needed for a square meter would cost about $2, which would be fully acceptable. The problem arises in matching indium requirements with indium production. Today's world production of indium is about 100 metric tons. Assuming an efficiency of 15% for the modules, about 7×10^6 m^2 of modules, needing 50 metric tons of indium, would be used to make a billion watts (a gigawatt) of PV capacity. This is equivalent to half of a nuclear power plant—not a great deal from a global perspective.

Three advances could lower this indium requirement by the time such a CIS production occurs. There is no theoretical reason opposing the use of thinner CIS layers. In fact, CIS is the most light absorbing semiconductor in PV, so CIS should make the thinnest cells. Layers as thin as 0.5 micron (or less) could be successful. Similarly, recycling of indium during manufacture would require rather simple procedures. These two advances would lower indium requirements to 1 gram/m^2. Then a gigawatt would require only 7 metric tons (rather than 50 MT). Finally, gallium is being substituted for about 5–25% of the indium in existing cells, and such substitutions could reach 50%. With all three of these changes, indium needs could be extremely small—under 4 metric tons.

Were the world's indium production to remain at 100 metric tons, the future of CIS would depend on the preceding approaches being developed. However, estimates of the concentration of indium in the Earth's crust show that it is present in a similar level as silver (i.e., about 0.05 part per million in the continental crust). Silver is mined at over 8000 metric tons annually. If indium can attain anything approaching this level, neither the amount of indium, nor its cost, would be a problem. In the meantime, large fluctuations in the price of indium could affect CIS. A tripling of indium prices would raise its cost contribution to $6/m², given existing layer thicknesses and material utilization. Obviously, these indium price fluctuations could be ameliorated by making the CIS layer thinner or raising the utilization rate of indium. Indium is a concern with CIS; but given the number of ways to reduce its impact, it is not likely to be a make-or-break issue.

Safety

Safety is another issue that becomes more important as large-scale commercialization of CIS becomes probable. Like other thin films, CIS has some serious safety issues associated with its manufacture. In this case, the issue is the use of hydrogen selenide during selenization. Hydrogen selenide is an extremely toxic gas. It presents risks very similar to those posed by silane or germane in a-Si manufacture. Manufacturers are confident of designing their facilities to use each of these gases safely. In fact, advanced designs for scaling up CIS manufacture call for making hydrogen selenide on-site by combining chemicals containing hydrogen and selenium. Only a small amount of hydrogen selenide would be stored for immediate use. Safety hazards would thereby be minimized. No hydrogen selenide would be transported from off-site for use. Another possibility is that hydrogen selenide will be eliminated and selenium itself will be used for selenization.

Even though there are several good avenues for handling toxic gases like hydrogen selenide, the question of toxic gases is a sobering one for an industry that prides itself on being an environmentally benign alternative. When the serious front-end safety issues related to making most PV modules become generally known, people will no doubt have to come to terms with the fact that even PV is not a simon-pure alternative to conventional energy sources. The author believes that they will decide that PV is still much better for the environment than other energy options.

Infrastructure

A final issue is the fragility of the US infrastructure supporting CIS. Although new companies—Solarex and Chronar—are starting work in CIS, no company, including ISET and Boeing, closely approaches the expertise of ARCO Solar. None besides ARCO Solar has even built a square-foot module, not to mention achieved record module efficiencies. Few except ISET can claim to have a manufacturable approach. Thus, the technology that may be considered the most promising new PV option is supported by only one key PV company. Globally, the situation is the same; no non-US company has launched a serious CIS effort.

During 1989, ARCO Solar was for sale. Speculation of its fate included sale to a foreign company, liquidation, or fragmentation into smaller companies. Uncertainty about the fate of ARCO Solar brought the progress of CIS into question. It also demonstrated the tenuous position of PV in general.

In August 1989, Atlantic Richfield signed a letter of intent to sell its ARCO Solar subsidiary to a giant West German electronics company called Siemens. By this action, Atlantic Richfield agreed to sell, for a small amount of cash (purported to be $30 million), the world's leading PV company, both in terms of annual sales and technological position. A decade of effort by those in the CIS technology was undercut, at least as far as the US status was concerned. With the sale's culmination (in 1990), this nation went from the global leader in CIS to a distant second. This sudden shift in fortune is symbolic of the entire US PV effort in the 1980s: despite strong technical progress, PV has gained little notoriety or support in our own country. Meanwhile, foreign investors—mostly German and Japanese—have been able to buy from the US what their own technical infrastructure did not invent: the best PV technologies in the world.

Recent discussions in Europe involving Siemens and a utility company called Bayernwerke in Bayerne, West Germany, have centered around the possibility of building a thin-film PV manufacturing facility on a site previously to be used for a nuclear reprocessing plant. Speculation exists that with Siemens' purchase of ARCO Solar (now called Siemens Solar Industries), the technology to be scaled up to this size is to be CIS.

Future of CIS

The CIS technology has established new performance standards for thin films. Also, at long last—after two decades of research—it has established the effectiveness of the thin-film approach. Until ARCO Solar's high efficiencies, the 15% module efficiency goal for thin films, established as a long-term goal by DOE, was considered nearly unapproachable. CIS has changed that. Today's CIS technology should allow someone (perhaps Siemens Solar Industries) to sell stable, 10%-efficient, low-cost modules within the next three to five years. With continued progress, the 15%-efficiency goal should be achieved, and probably before the end of the century. CIS modules of 15% efficiency costing $50/m² would produce intermittent AC electricity in the southwestern US at the impressively low cost of 4.3 cents/kWh; in New York, the cost would still be below 6 cents/kWh.

With the nearer-term technology (10% modules at $100/m²), the cost of PV power would be about 11 cents/kWh—a very competitive price for electricity available on summer afternoons. At that point—which will be reached when *existing CIS technology is manufactured on a reasonable scale* (around 1993–1995)—PV in general should become self-sustaining. No doubts will remain about the ultimate viability or usefulness of transforming sunlight directly into electricity.

11 ☼ Cadmium Telluride

The dark horse in the race for leadership among the thin films is cadmium telluride (CdTe). Today, it lags behind the others in performance. But it may have the highest potential, long term, for low-cost manufacture and for high efficiency. As such, it provides an additional avenue (along with CIS) to very low cost, and progress in cadmium telluride bolsters the overall probability that PV will meet its goals.

Interest in the material dates back almost as far as copper sulfide—about thirty years. Scientists have long recognized the importance of cadmium telluride because it has an almost perfect band gap (1.5 eV). This band gap is at the maximum in terms of its potential for high efficiency as a single-junction device. A polycrystalline thin-film CdTe cell was reported as early as 1972 by a German team, D. Bonnet and H. Rabinhorst, and interest in CdTe continued to grow, peaking in the late 1970s and early 1980s. Then interest seemed to dwindle until about 1987. At that point, corporate interest in CdTe accelerated again. Today, CdTe is in the third position among thin films but has one potentially crucial advantage—it may ultimately be the most inexpensive to make. CdTe is an unusual material in that it can be made well by a variety of low-cost methods. Almost any means of getting the right amount of Cd and Te on a suitable substrate can be used to make high-performance CdTe devices. The ease with which CdTe can be made challenges researchers to come up with the lowest cost processes for making it.

On the other hand, CdTe has some unique disadvantages, too. The worst is the presence of a heavy metal—cadmium—that is perceived as a serious health and environmental risk. However, the small amounts of cadmium present in CdTe modules should allow this problem to be managed.

Properties and Device Structure

Like CIS and a-Si, CdTe is a strongly light-absorbent semiconductor. Only about a micron is needed to absorb 90% of the solar spectrum. In fact, several key properties of CdTe have determined the structure that has been adopted for the best CdTe cells:

1. There are as yet no identifiable *high-band-gap* (over 2.4 eV), *p-type* semiconductors that can make good electric fields with n-type CdTe. Therefore, *p-type* CdTe has been used almost exclusively as the absorber material in CdTe cells, matched with a high-band-gap n-type material. The usual heterojunction partner is n-CdS (2.45 eV band gap).

2. As stated in #1, one must use p-CdTe as the absorber. But polycrystalline CdTe is difficult to dope highly p-type. This is not particularly a problem in terms of making a good junction (fairly resistive material works well, making a wide drift field), but it becomes a serious problem in terms of making good contact to p-CdTe. Good contact requires a highly p-type material in order to make a narrow, tunnel junction. Because p^+-CdTe is hard to make, contacting has always been a big issue with CdTe cells.

3. Cadmium and tellurium react exothermically (give off energy when they combine chemically) and form a CdTe compound that is almost always nearly stoichiometric, i.e., there is almost exactly one Cd atom for every Te atom. Extra Cd or Te are driven off during the reaction. The more subtle properties of CdTe (e.g., the number of free carriers, which can depend on a Cd or Te excess as small as one billionth of a percent) can then be adjusted by a heat treatment. The heat treatment characteristically is done at about 400° C in one of various ambients and chemicals, e.g., in air and cadmium chloride. Under these circumstances, the CdTe grains regrow into larger grains (Figure 31) that have fewer defects at their boundaries. At the same time, the number of free carriers can be changed, usually with the goal of increasing the number of holes and reducing the number of free electrons. This makes the material more p-type and freer of defects. The heat treatment is a critical step in almost all procedures used to

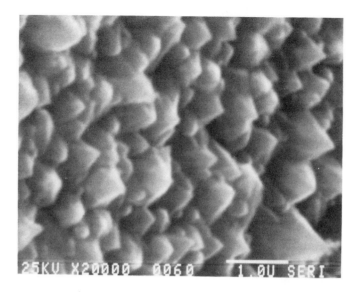

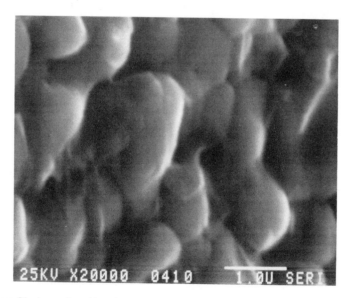

Figure 31. A postdeposition thermochemical treatment is usually critical to the performance of CdTe devices. Before heat treatment (top), exposed CdTe grains are small and conical; after treatment (bottom), grains enlarge and grow rounder. Defects at the grain boundaries are passivated during grain regrowth. (Courtesy of Ametek Inc. and SERI.)

make high-quality CdTe. In fact, most groups that are successful with CdTe find that the heat treatment is more critical than the CdTe deposition process.

4. Despite difficulties (see #2), several methods of making good contact to p-CdTe *have* been developed. Most depend on the heat treatment of #3 and a subsequent chemical treatment on an exposed p-CdTe surface. All are sensitive to any further processing. Any high-temperature process or any subsequent semiconductor deposition is likely to ruin the contact or affect its durability.

Because of these properties, almost all CdTe cells are now grown in a manner that leaves the contacting step for last. This means that no high-temperature step follows the contacting. But to do this, a particular cell structure has generally been adopted: glass/transparent contact (usually tin oxide)/n-CdS/p-CdTe/contact/metal (Figure 32). This structure mimics the a-Si structure in that the top glass is used as a starting material (also called a superstrate) upon which the layers are then sequentially deposited.

For various reasons, operational outdoor stability of CdTe modules has not yet been documented. Some CdTe devices degrade when they are exposed to humidity. The problem seems to be the back contact to the p-CdTe. Several approaches to making stable CdTe

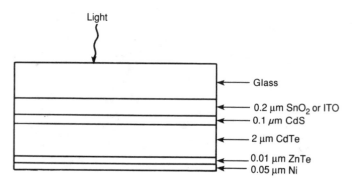

Figure 32. A typical CdTe cell starts with a glass superstrate on which conductive tin oxide or indium tin oxide (ITO) is deposited; then the junction is formed between a thin n-type CdS and the p-type CdTe absorber. An intermediate contact material such as p-ZnTe is optional. The final layer is a metal, usually nickel or graphite with copper.

modules are being developed, but the stability issue has not yet been thoroughly resolved.

The Easy Part: Making CdTe

As already stated, cadmium telluride seems to challenge the technologist to find ways to make it that are as simple and inexpensive as possible. It seems that no reasonable method has ever failed to make quality CdTe. New groups that have started to make CdTe cells have inevitably made rapid progress, usually making 10% cells within two years of initiating work—in great contrast to experience in other thin films, including a-Si and CIS.

In the late 1970s, four companies—ranging from some of the biggest to the smallest in the world—were developing CdTe: Ametek Inc. (Harleysville, PA), Monosolar (Inglewood, CA), Kodak (Rochester, NY), and Matsushita (Japan). Two of them (Ametek and Monosolar) were making CdTe by a method called electrodeposition, Kodak used a form of evaporation, and Matsushita used screen printing.

Electrodeposition

Electrodeposition is a method by which a material is deposited from a solution (called an electrolyte) onto a desired substrate during the flow of an electric current through the electrolyte. Apparatus can be as simple as a beaker containing the proper solution and two electrodes (one the desired substrate) suspended within the beaker to carry the current. When a voltage is applied, one electrode carries electrons to the solution; the other carries them away from it. In normal current flow through a wire, electrons carry the current. But within the electrolyte, positively and negatively charged ions (atoms lacking an electron or having an extra one, respectively) can carry the current. When they flow from the solution to the electrode, ions deposit on the electrode, forming the desired layer.

For the electrodeposition of CdTe, one electrode, called the cathode, is fashioned so that it is actually the substrate where one wants to form the CdTe layer. That substrate is usually glass/tin oxide/CdS. A metal clamp is applied to the conductive layers, i.e., to the tin oxide.

During electrodeposition, electrons flow into the cathode from the external source and pull Cd^+ atoms onto the glass/tin oxide/n-CdS cathode. At the same time, Te from the electrolyte is also being deposited as it reacts with the Cd to form CdTe.

Electrodeposition is an inherently low-cost process in the sense that the equipment needed to accomplish it is very inexpensive. It is also a familiar technology, used in industry to make coatings such as chrome on car bumpers. The only drawback of electrodeposition of CdTe is that it is slow—a micron can take an hour to deposit. But because the equipment is so inexpensive, cost per unit of throughput remains low. Additionally, as layer thicknesses drop—and that is the trend in CdTe devices—electrodeposition's slow rate may yet come to be an advantage, since it allows good control of ultrathin layers.

A little-known advantage of the electrodeposition of CdTe is its very efficient use of starting materials. Feedstock liquids containing Cd and Te can be added continuously for long periods without waste. Almost every atom of cadmium and tellurium added to the electrolyte ends up being deposited. None is lost because deposition depends on the flow of current, and this only occurs through the electrode—the desired substrate. Even when the electrolyte *must* be replaced (for instance, for periodical maintenance), it represents only a small fraction of the total material being processed. Even this may not be lost or require disposal. The waste electrolyte can be moved to another electrodeposition apparatus, and nearly all of the remaining Cd or Te can be electrodeposited and ultimately reused. This cleaning method—electrodeposition from a liquid—is actually the commercial means by which Cd is purified, so it is a familiar process. Absolutely minimal amounts of Cd or Te ever need to leave the manufacturing site for disposal. This is a clear advantage of the electrolytic method.

In the late 1970s and early 1980s, both Ametek and Monosolar adopted the technology of electrodeposition to make efficient CdTe cells. This meant that they had to master the subtleties of depositing compound semiconductors rather than simply depositing elements, as had been done previously. They succeeded, and in the process they patented most of the key procedures for making CdTe by electrodeposition. They also developed methods of optimizing CdTe devices, e.g., a 400° C heat treatment leading to grain regrowth and thermochemical processes used to contact the p-CdTe.

Some of the early work at Monosolar was funded through the

Department of Energy and SERI. But in 1983, the company was purchased by SOHIO, which then began work in CdTe electrodeposition at its Cleveland research center. But in 1984, British Petroleum (BP) bought SOHIO. Both Monosolar and SOHIO closed out their work in CdTe in favor of an expanded effort at BP in Middlesex, England. Although this shift represented an actual increase in effort from an international standpoint, it did mark a setback for the US in CdTe since work in Cleveland was abrogated.

Today, both Ametek and BP continue to have a major presence in CdTe. After some slowing, while BP transferred the Monosolar technology to England, the BP group is now moving very aggressively and is claiming impressive results. Among those *claims* (unverified) are the ability to make *stable*, one-square-foot panels of 9% efficiency. This achievement would put them second only to ARCO Solar (at 11% for CIS) in making efficient thin-film modules. They also claim to have achieved cell efficiencies of over 12% via electrodeposition, which would put them among the leaders in CdTe cells. BP Solar is a commercial distributor of silicon modules; they may soon add CdTe panels to their product line. Unfortunately, none of the claims attributed to BP have been confirmed by published results; nor have their modules been independently tested for stability. Like other oil companies, BP has been traditionally very secretive about their internally funded research progress.

The Ametek effort continued throughout 1989, but at a less ambitious level. Their development of modules—so crucial to thin-film commercialization—was slow. As such, Ametek was not a major industrial force. But Ametek has some of the best CdTe cells and had made several recent contributions to the technology (in addition to their pioneering work in electrodeposition). Among their patents were (1) a thermochemical treatment to optimize the CdTe film by grain regrowth, (2) a chemical treatment of the exposed CdTe surface to make a good contact, and (3) a design developed to resolve the contacting/stability problem by avoiding all contact to p-CdTe entirely.

This latter is called the n–i–p design, and it evolved in a rather unusual manner. Early work at Ametek revolved around making cells using n-type CdTe. This was based on the valid observation that n-CdTe was an easier material to make highly conductive and to contact than was p-CdTe. To this day, it would probably be preferable to make cells from n-type CdTe if it were possible to find a highly transparent, highly conductive *p-type* heterojunction partner for it. Unfortunately, no one has ever found such a material, and making very efficient cells with n-CdTe has never been done. As a compromise, Ametek matched n-CdTe with very thin *metal* layers. Like semiconductors, certain metals can induce an electric field when they touch a semiconductor. And metals that are thin enough (under 0.01 micron) can be quite transparent. Unfortunately, metal/semiconductor electric fields tend to be weaker than those induced between two semiconductors. Devices tend to be quite limited in voltage. After much work on metal/n-CdTe devices, Ametek researchers were forced to acknowledge that metal/n-CdTe cells would never be much more than 10% efficient. They abandoned their approach in the mid-1980s and took up the more conventional n-CdS/p-CdTe heterojunction structure. Their work on n-CdTe cost them dearly in terms of relative position among the CdTe groups and in terms of internal management support.

But one advantage was derived from their work with n-CdTe. Ametek had become familiar with a new material, p-zinc telluride (ZnTe). They had tried to use p-ZnTe to make good junctions with n-CdTe, since ZnTe has a reasonably high band gap (2.2 eV) and can be strongly p-type. But the band gap of ZnTe was still too low to allow much light to reach the CdTe, so it did not lead to efficient cells. However, it did allow Ametek to develop an entirely new design, a so-called n–i–p structure like that which is used in the a-Si technology. The structure they adopted was n-CdS/i-CdTe/p-ZnTe. Although this design was theoretically considered by others (including SERI) prior to Ametek's work, Ametek was the first to actually use p-ZnTe.

The middle layer in the Ametek n–i–p cell, intrinsic CdTe, was a highly resistive CdTe that was very slightly p-type. The Ametek n–i–p design used this i-CdTe in a way that avoided the problem of contacting p-CdTe. The structure, n-CdS/i-CdTe/p-ZnTe, induced a drift field across the entire i-CdTe layer in a manner analogous to the n–i–p structure in a-Si. Note that the Ametek approach avoided contacting the p-CdTe. Instead, contacting was to the p-ZnTe, and there was no problem contacting this highly conductive material. Ametek has published several stability tests of their n–i–p CdTe cells and small-area (100 cm^2) CdTe submodules. The test results are quite promising. Cells exposed to light for 3000 hours have been stable; as were small submodules exposed for the same period. As a further demonstration of ruggedness, some submodules were actually submerged underwater for a week. These also passed without degradation. Of all the designs for CdTe devices, the Ametek approach seems to have the best chance of being stable, because—unlike the others—it avoids contacting p-CdTe entirely.

Kodak and Matsushita Electric

Two electronics giants were also pioneers in the development of CdTe: Kodak and Matsushita Electric. Each made CdTe with their own unique, potentially low-cost method. Kodak's method, called close-spaced sublimation (CSS), was an evaporation method. First a suitable source material of CdTe was synthesized, usually by mixing and then reacting equal amounts of Cd and Te. The synthesized CdTe was then placed in a crucible and sublimated (a form of evaporation) in very close proximity (3 millimeters) of a substrate. During sublimation, the CdTe molecules moved from the source material to the substrate, depositing a thin film of CdTe. The substrate was glass/conductive oxide/n-CdS. Sublimation of CdTe was done at high substrate temperatures—about 550–600° C—and was very rapid: rates of microns per minute were achieved.

Kodak made rapid progress in CdTe: they announced a 10.5% cell in 1982, one of the first thin-film devices to achieve this milestone. (It was not independently confirmed under standard conditions.) Within another year they had made a 32-cm^2 submodule at 8.5% efficiency, perhaps the most efficient thin-film device of its size at that time. However, despite these successes, their work in CdTe was abandoned in the mid-1980s during a companywide reorganization. The reorganization was driven by the desire to cut costs by eliminating business opportunites that were not deemed directly related to Kodak's main business—photography.

The other electronics giant, Matsushita, has maintained its effort to the present. They are the only Japanese company to have a significant effort in any nonsilicon PV technology. Their CdTe work began in the late 1970s. They developed a uniquely inexpensive process—screen printing—but one with several severe drawbacks. Screen printing is a method in which a CdTe-containing paste is drawn over the surface of a substrate. The paste, whose function is merely to carry the CdTe material, is pressed onto the substrate, depositing a relatively thick (7 microns or more) layer. Layer thickness is one of the disadvantages, since it increases materials costs and does not allow flexibility in terms of newer designs that depend on thinner layers.

A thermal treatment called sintering follows screen printing. The printed paste is heated to over 600° C, which drives off the paste's non-CdTe material and leaves relatively good CdTe behind.

Because of the high temperatures attained during sintering, Matsushita has been unable to use tin oxide, which degrades under such thermal stress. Instead, they adopted a glass/n-CdS/p-CdTe design. But this design has been limited by the fact that CdS cannot be made as conductive as tin oxide. Conductivity is crucial in the top layer of a cell, since current flows laterally across this layer to reach metal interconnects. With tin oxide as the top layer, cells can be relatively wide (about 1 cm) with only minor resistance losses. With the less conductive CdS, cells must be much narrower, or they would experience prohibitive resistance losses. Matsushita has been forced to adopt a module design with many very narrow cells. As a result, their design has a great deal of inactive area—area lost to interconnects that produce no power. The ratio of the active-to-total area of Matsushita modules is only 60%, the lowest by far in the thin-film industry.

Because of their poor active area ratio, Matsushita is capable of making very efficient cells but very *inefficient* modules. They have claimed cell efficiencies above 10% since about 1983. But their modules have remained at or below 6% efficiency to this date. Even if their module efficiency measured on an active area basis is 10%, the total-area efficiency of their modules is only 6%. Matsushita has had this problem for almost a decade without resolving it. However, they have made contributions in two other areas: consumer products and module stability.

CdTe cells made by Matsushita have been used in calculators made by Texas Instruments for several years. They are remarkably good cells for this application, being very efficient even under poor illumination. They are the only CdTe devices (and the only nonsilicon PV devices) in commercial use (although CdTe and CIS modules are soon to become available from other manufacturers). Meanwhile, Matsushita has reported testing their CdTe panels outdoors with good results. Matsushita uses a simple graphite paste as their contacting material. The paste has an additive—copper—to improve its contacting properties. Careful sealing of the modules to prevent the ingress of water or water vapor is an essential part of their module encapsulation approach. Matsushita has reported that CdTe modules tested outdoors for up to four years have been stable. But these results, like the stability claims of BP, have not been independently confirmed.

Plateau Period

The mid-1980s was a plateau period for CdTe. Kodak dropped out. Monosolar was swallowed up by SOHIO, which in turn became subsumed under BP. The latter did not publicize their progress. Ametek shifted to a new approach and retrenched. Matsushita achieved 5–6% square-foot modules but did not commercialize or improve them. One other giant company, Atlantic Richfield, in the form of ARCO Solar, entered and then left the field precipitously. At ARCO Solar, they used the Kodak approach (CSS) to rapidly and easily achieve 10% cell efficiency. But this progress occurred at the same time that a-Si and CIS were also doing well at ARCO Solar. In a reorganizational move, they chose between CIS and CdTe, eliminating the latter. Their decision seems to have been based on problems with CdTe contacts as well as the presence of Cd in the CdTe technology.

By the mid-1980s, over ten groups (Monosolar, SOHIO, BP, Ametek, Kodak, ARCO Solar, Matsushita, and a university—Southern Methodist University) had achieved over 10% efficiency on cells, but the technology appeared to be going nowhere (if one excepts BP, which was very quiet about their progress). That apparent stasis changed with developments at a tiny company in El Paso, Texas: Photon Energy.

Photon Energy

Photon Energy is an interesting company which mirrors the progress of PV in general. "Resource limitation" and "Photon Energy" are synonymous. At this writing, Photon Energy has 11 employees, yet it has accomplished some of the most impressive achievements in the progress of PV.

Photon Energy was the resurrection of an earlier company—Photon Power—presided over by a pioneer in thin-film PV—John Jordan. Photon Power made great strides in developing low-cost manufacturing of copper sulfide modules, but failed—with everyone else in that technology—because of the material's instability. A few years later, Jordan purchased the shell of Photon Power and shifted it to work on CdTe. Photon Energy, as the new company was called, used the same *modus operandi* as Jordan had before, emphasizing extremely low-cost processes and product-sized devices—1-ft² modules rather than 1-cm² laboratory cells. In just a few years, Photon Energy was able to make state-of-the-art CdTe cells and modules. Then, in 1987, they successfully competed for a SERI award and were funded for modest amounts (about \$250,000/year) through 1989. However, for them, these modest funds were a substantial part of their total R&D effort, and they leveraged them (in terms of results) as if they were millions.

In their first three years under SERI contract, Photon Energy became the acknowledged US leader in CdTe. In 1988, they made a 7.3%-efficient, 1-ft² panel (verified at SERI), thus surpassing all others in the field. (BP's claims of 9% have never been made officially or confirmed independently.) The active-to-total-area ratio of the Photon Energy module was 90%, showing that the Matsushita problem (60% active area) was not generic to CdTe. Meanwhile, Photon Energy also developed an innovative device structure in which they were able to achieve 12.3% efficiency (also SERI verified) on a small cell from one of their submodules.

Their advanced cell had an improved response to high-energy photons. It was the CdTe analog of the ARCO Solar CIS design that used thin CdS to accomplish the same improved high-energy response. As such, it was the first high-efficiency cell in the CdTe technology to accomplish this. The concept—straightforward in idea—is difficult in practice using the conventional CdTe structure in which CdS is the second layer deposited. The CdS is subsequently exposed to several high-temperature steps. But Photon Energy found a way around this problem in their world-record CdTe cell. At this writing, they have yet to incorporate the advanced cell design in their larger modules. When they do, the new design should result in an immediate increase in module efficiency to over 9%.

But the real strength of Photon Energy is in their ability to make almost all the layers of the CdTe module using low-cost processes. They have found high-rate, nonvacuum methods for all of their semi-conductor layers. Many of their methods are similar to those that they developed during their work in copper sulfide, e.g., spraying methods in which suitable solutions are atomized and deposited on a moving substrate. Like electrodeposition, their methods are highly efficient in the use of material: more than 90% ends up in the module, and almost none is exhausted as either liquids or fumes.

The low costs of the methods used by Photon Energy to make CdTe panels reflect a true concern to reach the cost potential of thin films. In their projections for the near term, Photon Energy expects to make CdTe modules at costs very close to $50/m^2 without need to scale their production beyond about 6 megawatts of modules per year. This is the lowest price and smallest initial production requirement of any in the industry, giving Photon Energy a special flexibility in terms of market response and capital requirements. At their current 7% efficiency and near-term cost of about $70/m^2, Photon Energy should be making modules at an (up until now) unheard-of cost of only $1/W_p$. With expected improvements, their costs should drop substantially below that figure. They have the potential to make both the efficiency and cost goals of the DOE, which would result in modules costing as little as $0.33/W_p$ (15% efficiency at under $50/m^2).

Photon Energy has begun the process of commercializing CdTe panels. A consortium of utilities and DOE has agreed to purchase 20 kW of their CdTe panels for a demonstration in 1990. This will be the first such major use of CdTe panels. Photon Energy has scaled up their design to 4-ft^2 panels to meet this purchase. Their main barrier—the same as for everyone else in the technology—is achieving stability. As yet, Photon Energy has not settled on a module encapsulation approach. They need very good isolation from the environment to minimize the invasion of water vapor, a critical requirement for outdoor stability. The instability of CdTe is of a different character than the instability of copper sulfide—and far less in overall magnitude. Yet Photon Power failed because of stability issues. The same problem, now with a new material, haunts its offspring, Photon Energy. The good news is that recent tests at SERI (1990) show that Photon Energy CdTe modules are stable for over 300 days (the period of the test).

Issues

Several issues cloud the future of the CdTe technology. Like any other thin-film technology, CdTe needs to improve in efficiency, but this is actually a secondary matter. Cells have reached 12%, and the potential for much higher efficiency—as high as for CIS, i.e., 20%—is a reasonable possibility. Unlike a-Si or CIS, very low-cost processes for almost every step of making CdTe modules have been developed. So the usual issues—efficiency and cost—seem relatively minor. Indeed, at $50/m² and 15% module efficiency ($0.33/$W_p$), which are reasonable goals, the CdTe technology will be producing intermittent AC electricity in the Southwest at about 4.3 cents/kWh.

On the other hand, stability remains a thorny problem. Both Matsushita and BP claim to have achieved it, but without independent verification. Ametek seems to have a better approach (contacting p-ZnTe instead of p-CdTe) but they are far from demonstrating it outdoors with a module/product. Photon Energy is a tiny company with big plans and big challenges: its ability to reach an optimal encapsulating scheme in a very short time seems risky. Until stability can be demonstrated, the CdTe technology will not be fully appreciated, nor is it likely to reach full commercial potential.

Another issue of quite a different sort hangs over the technology: cadmium as a toxic material. The presence of cadmium is a different issue than for the other thin films, where almost all environmental and safety problems are confined to the manufacturing phase and do not leave the plant with the product. Cadmium is part of CdTe modules. It is a heavy metal that is despised in Japan because of a public health catastrophe in the 1950s when cadmium was blamed for an outbreak of a bone disease called Itai Itai (after the town in which it occurred). In the US and Europe, cadmium is feared but tolerated. Still, the introduction of a new, major use of Cd—even if it is arguably a safe use that offsets much more noxious methods of making energy—may raise serious opposition.

Cadmium Risks

How much cadmium is in a thin-film cadmium telluride panel? In full production, such panels should have less than 10 grams per square meter (g/m^2), and possibly under 3 g/m^2. Assuming about 10,000 m^2 per megawatt of PV capacity (10% efficiency), between 30 and 100 kg of Cd will be in a megawatt field of CdTe panels. These are not large amounts of Cd from the standpoint of familiar uses. Over 20,000 *metric tons* of Cd are already used every year for such things as coatings on metal, batteries, red and yellow paints, treated vinyl, and plasticizers. A Ni/Cd, size-C battery has about 20 times more cadmium in it per watt of capacity than does a CdTe module, although there is no warning or statement about quantity of Cd contents on the battery package. And cadmium is present in very large amounts as a by-product of zinc production, in coal fly ash from coal-burning plants, and in phosphate fertilizers (3–120 mg/kg). The latter is a serious health threat in that phosphate fertilizers are used ubiquitously and provide a direct route for cadmium to enter the food chain of humans through vegetables and grazing animals.

For PV, cadmium is an issue in two senses:

□ Workplace safety
□ Product use and disposal

But manufacture is an area in which Cd poses few significant barriers. The two leading methods of making CdTe—electrodeposition and spraying—are both very efficient in their Cd use. Well under 10% is wasted in spraying and less than 1% in electrodeposition, and in both the cadmium can ultimately be gathered as a solid waste, which is easily handled. The waste in spraying is excess solid CdTe on the edges of the panel; for electrodeposition, there is negligible waste because all the effluent electrolyte can be pretreated to remove the cadmium.

Actually, there are advantages to the CdTe production process in comparison to other PV facilities. Cadmium poses a *chronic* rather than toxic threat. It takes time and many exposures for lungs, bones, or kidneys to be affected by cadmium poisoning. Careful monitoring of equipment and of employees (urine samples) should allow a CdTe company to minimize accidental exposures long before they become a problem. In this sense, working with Cd, as opposed to working with other thin films with their toxic gases (silane, diborane, phosphine, germane for a-Si; hydrogen selenide for CIS), is actually much easier and less dangerous. In those other cases, one exposure could cause immediate death.

Product use and disposal are another matter. Laws affecting Cd-containing products are expected to tighten significantly. The real question is: Do CdTe panels pose a threat to our health and environment?

The biggest product-use concern about CdTe modules seems to be related to house fires and fires in commercial buildings. CdTe is a very stable compound that melts at a high temperature (over 1100° C). Temperatures attained in a house fire are usually not high enough to vaporize CdTe. However, even if CdTe does vaporize, the amounts in a CdTe panel are so small—say 200 g in total for a 20-m^2 array—that safety limits would almost never be exceeded. Meanwhile, other threats from house fires—e.g., polyurethane in sofas—are far more likely to be life-threatening.

Product disposal is another concern. An encapsulated panel is designed to keep out the environment. It is also good at keeping the CdTe from entering the environment. Experiments are now under way to determine if a CdTe module would actually leach Cd into groundwater if it were buried in an uncontrolled municipal landfill. Leakage is unlikely.

Even if it is found that CdTe panels are not threats to groundwater, CdTe manufacturers might want to minimize public concern by other strategies. For instance, CdTe panels could be disposed of the same way other hazards are now handled, in a controlled landfill for toxic wastes; or the CdTe panels could be recycled by feeding the panels directly into a cadmium smelter. The highly pure cadmium and tellurium could then be sold back to the PV company as source material for another generation of CdTe modules. This approach makes sense to the resource companies involved because they recognize the need for cradle-to-grave control of cadmium. Recycling could be cost-effective as well, because of the value of the purified cadmium and tellurium.

At this point, it is hard to tell whether Cd in CdTe is an issue or merely a tempest in a teapot. Other Cd-containing products already exist with little real impact. A child has a far greater chance of ingesting Cd from a radish grown in phosphate fertilizer than he has of being endangered by a panel on his roof. In a fire, the polyurethane in the sofa has a far higher probability of killing you than a roof panel that is in itself almost too stable to be threatened by the temperatures of a house fire.

The threat of Cd is probably miniscule compared to the advantages of offsetting other energy production methods. Further study of these issues should help quantify the real trade-offs. But for perspective, let us take a look at an initial analysis that has been done on a very familiar but almost unknown source of inadvertent cadmium emissions, coal plants.

Coal has several heavy metals or hazardous materials in it, including cadmium, arsenic, selenium, and tellurium. In fact, we can calculate the amount of these materials that a coal plant produces as waste on a per kilowatt-hour basis. It turns out that a coal plant produces cadmium at the rate of 1 g/MWh. This is the amount of cadmium that inadvertently turns up in coal fly ash or is disbursed through the stack. Coal fly ash is usually spread on uncontrolled landfills or used as a component of cement.

How much cadmium does a CdTe module use per megawatt-hour? If one assumes a 10% efficiency and 5 g/m^2 of Cd in a CdTe panel, a CdTe panel lasting 30 years outdoors in Kansas would require cadmium at a rate of 1 g/MWh. *This is the same as the inadvertent release of cadmium by a coal plant.* This 1 g/MWh is not the cadmium wasted during CdTe module manufacture (which is much less; say 1–10% of this amount). It is the *total cadmium* in a cadmium telluride panel. So instead of an inadvertent release, as in the coal plant, we have a controlled use of cadmium that is of the same order. Meanwhile, in addition to its inadvertent release of cadmium, the coal plant is producing 12 g/MWh of selenium, 120 g/MWh of arsenic, and plenty of better known pollutants such as sulfur dioxide, nitrous oxide, particulates, and carbon dioxide, among others. This is the kind of relative size of the cadmium issue for PV in relation to cadmium as a threat from other, less controlled sources.

In fact, CdTe panels may be part of the solution rather than the problem. Cadmium is inadvertently released from processes that are usually considered essential—e.g., zinc refining, coal plants, phosphate fertilizers. The owners of these facilities could be required to remove their cadmium and feed it into a controlled use like CdTe modules—in which the cadmium would then be sealed between glass sheets for 30 years, then recycled.

But the mere presence of Cd in CdTe assures that for those who may not be interested in quantifying trade-offs, CdTe will arouse opposition. The cadmium issue has often been cited as limiting by those PV companies and research organizations that do not work in the CdTe technology. How the technology fares—whether it is regulated or (in the extreme) is banned—will depend on how society views the age-old problem of balancing potential value against limited but unfamiliar risk.

12 ☀ Concentrators

We have completed our discussion of thin films. Concentrators are on the other extreme of the PV spectrum. Whereas thin films emphasize low cost at reasonable efficiencies, concentrators are based on reaching exotic, ultrahigh efficiencies while maintaining affordable manufacturing costs.

Concentrators are based on a clever idea for reducing the cost of PV electricity by using large lenses to focus sunlight on small cells. In this way, cell area is traded for less expensive lens area. At the same time, more expensive, and thus more efficient cells are affordable, because fewer of them are needed.

If a lens 1000 cm² in area focuses sunlight on a cell 1 cm² in area, the concentration ratio is 1000 × —which is also called 1000 suns, indicating the intensity of the sunlight on the cell versus that of normal sunlight. Concentration ratios are in the range of 2 to 1000 suns for different concentrating systems, although even 1000 suns is not a theoretical limit.

The concentration ratio is a critical parameter defining the economics of concentrators. Let us assume we have a system operating at 1000 suns, and each cell being used is 1 cm² in area. Then 10 cells would be required for each square meter of module area. One could afford to spend a lot more for those ten, 1-cm² cells than for 10,000 of them (as would be needed for a square meter of 1-cm² cells). This is the basis for savings associated with concentrators.

Concentrator Modules

Actually, it is not as simple as trading lens area for cell area. A lot of other hardware is associated with concentrators. Figure 8 (see page

33) shows a typical concentrating module. Such a module consists of a boxlike housing (the outer frame of the module), a set of lenses, cells at the focus of those lenses, a cell housing that includes an arrangement of mirrors to refocus light that would otherwise miss the cell, and electrical interconnections between the cells. In effect, besides lenses, one is adding a lot of other apparatus not required for nonconcentrating flat plates. Each of these has manufacturing and handling costs.

Another important aspect of concentrators is the need for cooling the cells. Without cooling, cells could reach temperatures as high as 1000° C under focused sunlight. Such high temperatures would destroy the cells. Even modest temperature increases—to above 90° C—would cause problems, because all cells naturally lose efficiency with increased temperature. To avoid any significant temperature rise, concentrator cells are individually bonded to a material capable of rapidly dissipating heat to the module housing. Often the housing is constructed with heat-dissipating metal fins on the back to rapidly move this thermal energy from the cells.

How well can heat be dissipated from a cell? A cell under a lens—by definition—is surrounded by an area without cells that is equal to the area of the lens. With good heat dissipation from the cell to the surrounding module area, cell temperatures can be kept close to the temperature of a 1-sun, flat-plate module. In effect, the same heat is being spread over the same area in both cases—as long as the passive cooling is done well enough. The trade-off is the cost of the heat dissipation mechanism.

General Characteristics

Several important characteristics distinguish concentrators from the flat-plate systems we have examined in previous chapters. Concentrator modules must track the sun closely to be effective. Systems with low concentration ratios (under 20) are not as sensitive to misalignments between the sun and the lenses and can afford single-axis tracking; but these low-concentration systems are not likely to be the most cost-effective. Those with high concentration ratios will be more economical but require very precise, state-of-the-art, two-axis trackers. If the tracker is not precisely directed at the sun, the lenses will not be able to focus the sunlight on the cell. Little or no electricity will be generated by a concentrator module that is not closely aligned with the sun.

But very precise trackers are more expensive than other trackers and have higher maintenance costs because they are more vulnerable to breakdown, especially from winds. Also, two-axis trackers require a large land area around them to allow full movement as they track the sun while also not shadowing other trackers. This increases land requirements about threefold in relation to a fixed, flat-plate array of the same efficiency.

Inability to Use Diffuse Sunlight

Perhaps the most crucial negative quality of concentrators is that they cannot use diffuse sunlight. Diffuse sunlight is the sunlight that we see as sky light. On a cloudless day, almost 20% of the sun's energy is diffuse. We see that light as blue sky. On a hazy day, or one with broken clouds, this percentage can be higher (30–40%); on a cloudy day, we may still receive 30–70% of the sunlight of a cloudless day, but all of it will be diffuse. Concentrators cannot use any of this diffuse light because they cannot focus light that is not perpendicular to the lenses.

The effect of this limitation is to make the economics of concentrators much less favorable outside of normally cloudless regions like the desert areas in much of the US Southwest. These areas have very high amounts of annual sunlight, and about 70–80% of it is not diffuse. Concentrators work very well in these very sunny regions. They work poorly anywhere else. In typical US locations outside the Southwest, almost 50% of the sunlight is diffuse. In these regions, flat plates are likely to dominate. Of course, even this apparently obvious observation is an oversimplification: if the cost of flat plates or concentrators differ enough, one or the other could still dominate outside of its natural solar region. In addition, system characteristics enter into the equation. If someday we have superconducting storage and transmission, or can make hydrogen with PV and transmit it over long distances via pipelines, using concentrators in the desert to supply distant urban regions might look especially attractive. Or, it might turn out that flat plates based on thin films are cheaper than concentrators, even in the desert.

Lenses and Other Losses

High-quality, low-cost lenses are essential to the success of concentrators. Glass lenses would be the best choice in terms of their transparency and optical perfection, but they are prohibitively expensive. Over the last decade, another lens—the plastic, Fresnel lens—has come to dominate. Fresnel lenses can be mass-produced at reasonable prices. However, they are far less optically perfect than glass lenses, focusing only about 85–91% of the sunlight. Their large losses are from light striking the raised edges of the Fresnel lens and from reflections from the top and bottom surfaces of the lens.

This means that even under the best circumstances, a concentrator starts with a great debit in relation to a flat-plate system. A minimum of 20% of the sunlight is unusable, diffuse light; and another 10–15% is lost because of the imperfections in the lens.

Economics

Although the apparent attraction of concentrators is to use fewer cells to reduce cost, the real advantage comes from using unique cells that would be prohibitively expensive if one had to use more than a few of them. But these cells can be by far the most efficient of any cells used in PV.

In fact, concentrator modules generally cost far more than flat plates—especially thin films; and they are also far more efficient. For instance, goals for flat-plate efficiencies tend to top off at 15%, with 20% module efficiency considered a very long-term possibility at best. But concentrator module efficiencies tend to start at 20% and go up—with 40% on the long-term horizon. We can speak with confidence of reaching costs as low as $50/m^2 for thin-film modules; for concentrators, three times that might be very hard to ever achieve, as we will see. And concentrators must use two-axis tracking, which also adds expense in relation to flat plates. Thus, the general picture, in terms of concentrator economics (especially in relation to thin films), is one of high cost but high efficiency.

To have a better understanding of the potential of concentrators, we must know something about their component costs. Figure 33 shows the relative costs of the different components making up a concentrator module. The costs are a projection based on technology that exists in the laboratory, but which has never been produced in any quantity. The projected module cost is about $200/m², and module efficiency is about 20%. This efficiency already takes into account lens losses; it actually assumes 22.5% cells. Cells as high as 28–37% have been reported, so (in terms of efficiency) this is a conservative projection of an existing technology. At what cost would such a system produce electricity?

Sandia National Laboratory says that two-axis tracking for concentrators now costs $200/m² (all area-related BOS costs). Let us assume we are in a desert, with 3.1 MWh/m² of sunlight a year available for our two-axis tracker. If we reduce this by 20% because that much is unusable, diffuse light; and 15% for system losses (temperature and wiring), i.e., by 35%, then about 2 MWh/m² remains. At 20% module efficiency, our system will produce about 400 kWh/m² annually. Thus, our DC cost would be about 12 cents/kWh$_{AC}$.

This is as good a cost level for a "near term" technology as any we have examined. Only the projections for CdTe and CIS compare well with it. A germane question is the actual cost of the cells that would be used in such a system. We assumed that 34% of the $200/m²

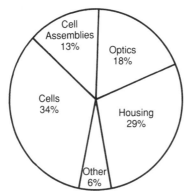

Figure 33. The major component costs of concentrator modules are cells (and their interconnects), the module housing, the cell and its supporting components, and the lenses and secondary focusing elements. This breakdown is approximate and is based on cell costs of about $70/m² and total module cost of about $200/m².

was cell cost (Figure 33). Thus, we are allowing about $70/m^2 for cells. How much is this for an individual cell? At 200 suns concentration, there would be 50 cm^2 of cells, and we could afford to pay about $1.40/cm^2. Can the existing high-efficiency cells be manufactured for this cost? If they cost $5/cm^2, they would contribute $250/m^2 instead of $70/m^2 and raise the cost of electricity to 17 cents/kWh.

We do not really know what the concentration ratio will be of an economical concentrator. Suppose our module was designed for 1000 suns rather than 200. Then we would have only 10 cm^2 of cells and could afford $7/cm^2 for them to keep the total cell cost at $70/m^2. This is an important point: cell economics depends crucially on the concentration ratio. The higher the concentration, the more we can afford to pay for individual cells.

Two cell technologies are competing to be used for concentrators, and they fall into two camps: low cost but low concentration versus higher concentration but higher cost. In the near term, the first kind—based on single-crystal silicon cells—may dominate; in the long term, more efficient cells (but more costly), based on single-crystal gallium arsenide, should displace them. We are already seeing evidence of this evolution in some astonishing efficiencies being achieved by gallium arsenide cells.

What kind of long-term cost and performance goals do concentrators need to reach in order to be competitive? Let us suppose that area-related cost for two-axis tracking comes down to $140/m^2. Sandia believes that this is possible. Let us compare goals based on *two* long-term module efficiencies, one for an assumed silicon technology, the other for gallium arsenide materials. Let us take these *module* efficiencies to be 25% and 35%, respectively. How much can we afford to pay for such modules if we want AC electricity at 4.5 cents/kWh?

Calculations show that for the lower efficiency assumption (25%), we can afford to pay $27/m^2 for our concentrator modules; for the higher efficiency, we can afford $105/m^2. The relatively low-efficiency approach based on silicon will never get us there (4.5 cents/kWh$_{AC}$) unless we are far off on our tracking costs. But the higher efficiency module allows a much higher module cost, but one that is still too low unless we change our assumptions about other cost components. After all, in previous calculations, we allowed $70/m^2 for the cost of cells *alone*. Now we can afford $106/m^2 for the whole module, including cells.

We said that near-term modules (without cells) should cost about $140/m². But if such modules are ever manufactured in great quantities, their costs could fall by 50%, i.e., to about $70/m². Then we could afford another $36/m² for the cells and meet our $106/m² total. At 1000 suns, we could afford to pay $3.60/cm² for 10 cm² of our high-efficiency gallium arsenide cells.

These are nothing but ballpark estimates. Suppose we could design cells for a *2000- or 10,000*-sun module? At 10,000 suns, we would have one 1-cm² cell per square meter and afford to pay $36 for it. Or suppose area-related costs of two-axis tracking fall to $100/m². We would be able to afford cells costing $76/m². Our options would be quite different. We must always bear this uncertainty in mind when making judgments between PV technologies: overdependence on simplified cost calculations (based on limited technical assumptions) can be hazardous to our perspective. Our estimates are only of use to give us some idea of the goals that concentrators should achieve to make economic sense.

Functions of Concentration Ratio

There are several other properties of concentrators (besides cost) that are functions of concentration ratio. One is quite fortuitous for all concentrators: cell efficiencies rise to some maximum as concentration ratio increases. Naturally, we expect cells to put out more power as illumination rises. But, interestingly, their *efficiencies* also rise with illumination, so they put out even more power than would be expected.

The physical reason for this increase in efficiency with illumination is somewhat complex. The cell's current goes up proportionally with the amount of illumination, as one expects. By itself, this would not result in increased efficiency. In fact, the major factor affecting improved efficiency is that cell *voltage* increases as current increases. Why?

We can use our analogy of a dike again. Voltage is analogous to the force it takes for water to rise above or penetrate a dike. If the height of a dam is low, or it has many leaks, the analogous voltage is low. Another way to put it is that the ratio of water kept back by the dike to the water it leaks defines voltage. The greater this ratio—the more effective the dike—the higher the voltage. Under normal sunlight (1 sun), the solar cell electric field (the dike) holds back a certain amount of water—i.e., electric current flows externally rather than back through the electric field. But under 100 suns, the same field results in 100 times more electric current. In both cases it leaks the same amount of water through a few holes (defects in the junction region that allow some electrons or holes to return to the side from which they came). In good cells—i.e., the kind used in concentrators (single-crystal devices)—the leaks do not change with illumination. Since voltage is the ratio of current to leaks, and current has increased 100-fold while leaks are unchanged, voltage increases.

The increase is actually fairly slow (in mathematical terms, the voltage is a logarithmic function of the ratio of current to leaks), but the improvement is not insubstantial. If cell temperature can be kept down (i.e., if passive cooling can be done effectively), cell efficiencies can rise significantly. For instance, a cell under 200 suns of illumination is likely to be quite a bit more efficient than the same cell measured at 1 sun (if the competing effect, the temperature, is held constant). A cell might go from 20% efficiency at 1 sun to 25% at 200 suns. This is a fairly typical improvement caused by the increase in voltage under focused sunlight.

Some of the properties that are a function of concentration ratio are loss mechanisms. One is temperature sensitivity. As we know, all cells lose voltage with temperature. However, the loss is inversely proportional to the cell's voltage, so higher voltage cells lose proportionally less. They maintain their efficiencies better. Desert locations are likely to be used for concentrators. The high temperatures resulting from less-than-perfect heat removal favor the use of higher voltage (higher band gap) materials such as gallium arsenide rather than low-band-gap materials like silicon.

Suppose we had perfect heat dissipation methods. Why wouldn't concentrator cells just keep going up in efficiency with an increased concentration ratio? Can't we just perfect higher and higher levels of concentration and thereby achieve extremely high efficiencies?

Besides the inadvertent increase in temperature at higher concentrations (perhaps a problem that can be controlled), there are two other competing effects that limit improvement and eventually degrade performance. An efficiency peak is achieved at some level of concentration because of these deleterious phenomena. The competing effects are (1) a complex phenomenon that has to do with the density of free carriers being produced by the concentrated sunlight and (2) increased losses associated with high series resistance. At high concentrations, huge numbers of free carriers are being generated by light. In fact, so many are produced that the numbers of majority and minority carriers tend to become similar on both sides of the junction. This is like the problem that all semiconductors face at high temperatures: the more similar the number of electrons and holes, the smaller the induced electric field. At the extreme, both materials have almost the same number of carriers and no field is induced. Voltage falls when differences in free carrier densities diminish. Like analogous temperature-dependent losses, the effect caused by concentrated light is also inversely proportional to voltage. It affects low-voltage materials like silicon far more than a higher band gap (higher voltage) material like gallium arsenide. The result is that gallium arsenide cells can be used under higher concentration ratios before their efficiencies start to peak. Silicon seems to be confined to under 500 suns; gallium arsenide can be used at concentrations approaching 1000 suns.

Series resistance in a cell is the resistance to current flow through the cell's various layers and contacts, as well as the resistance to flow from one layer to another. Total resistance losses are a function of the *square* of the number of electrons—the current—flowing in the cell. As such, losses are terribly sensitive to increases in cell current. But, as we know, concentrator cells are very high-current devices. Their current is proportional to the increased illumination level, i.e., to the concentration ratio. On a per square centimeter basis, they may have 200–1000 times the current of a 1-sun cell. Resistance levels that could be tolerated in a 1-sun cell would be intolerable at 100-sun levels and ridiculous at 1000 suns—resistance losses go up with the square of the increased illumination and would be higher by a factor of 10,000 at 100 suns and *1,000,000* at 1000 suns. This implies that concentrator cells must be specially made to have minimal resistance throughout their structure.

A generic result of the need to minimize resistance is that concentrator cells are usually made with very heavy metal front grids. They need a lot of metal to carry off currents as high as 10–30 amps/cm^2. The question then becomes, how much light is lost due to grid shadowing? At one time, this was a significant problem with concentrators. But in the last three years, a new device—called a prismatic cover glass—has been developed that seems to solve the problem and allows an important advance in concentrator cell efficiencies.

A prismatic cover glass is a gridwork of triangular, glass prisms that lie over the metal grid of a concentrator cell (Figure 34). Light striking the prisms, which would normally be lost because of the metal grid beneath them, is refracted to the side and into the cell. Cells designed to have prismatic cover glasses can be heavily gridded without major light losses, a significant step forward for concentrator technology in general. As with other components, prismatic covers will have to be inexpensive and durable if they are to succeed.

Concentrator Cells

So far, we have only obliquely alluded to the high-efficiency cells that go into concentrator modules. We know that they have to be as efficient as possible. (At only 15% efficiency, even a cell costing nothing would provide AC electricity at 9 cents/kWh with the best assumptions about the rest of the concentrator module. Higher efficiency cells have much better potentials.) In general, the most efficient PV cells are made from single-crystal materials. Besides the fact that they can be made from different materials, they also fall into two basic device

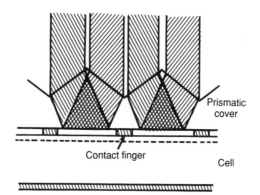

Figure 34. A prismatic cover can be designed to refract light away from contacts and onto the active area of a concentrator cell. Concentrator cells can now be designed for very dense grid coverage without incurring large shadowing losses.

categories: single- and multijunction cells. Under *1-sun* illumination, single-junction cells are limited to theoretical efficiencies of about 28%; and they fall off from this value as their band gaps vary from the optimal, 1.5 eV. But two-junction cells have the theoretical potential of about 35% efficiency under 1 sun; and three-junction cells have a potential of 41% under 1 sun. This latter is not a theoretical limit in terms of the ultimate potential of PV. Cells with *many* junctions (e.g., 6 or 7) could get to almost 50% efficiency.

Again, this is not the theoretical limit for PV. We know that cells *gain* efficiency under concentrated sunlight. The theoretical efficiencies quoted above for 1 sun are well below those appropriate for concentrator cells. We may expect about 33% single-junction cells and 44% two-junction cells under 1000 suns concentrations, assuming optimal band gaps and optimal designs. The theoretical limit, based on upwards of 10 junctions and 1000 suns illumination, would be 60% efficiency — and even this is not the final limit of PV cells (since higher concentrations may eventually become practical). It is important to realize that despite some very impressive results, we have hardly scratched the surface of the practical potential of multijunction cells for concentrators.

Thus, it is natural for us to separate our discussion of concentrator cells into two categories: single- and multijunction cells.

Single-Junction Cells

Two kinds of single-crystal cells dominate single-junction concentrators: silicon and gallium arsenide (GaAs). Of the two, the silicon cells are less expensive and less efficient. But they still dominate because currently they are more easily manufactured, especially in small volumes.

The near-term direction of research in concentrators is toward practical, 20%-efficient modules. Sandia National Laboratories announced the achievement of a prototype 20% concentrator module based on silicon cells in 1989. Two types of silicon cells seem to have the potential to get to 20% efficiency in manufactured modules: they are called the Green cell and the Swanson cell. (Both were covered in Chapter 7 as single-sun cells.) The Green cell (named after Australia's Martin Green) was used in the Sandia prototype 20% module.

The Green cell is a fully optimized, single-crystal silicon cell of the classical design: that is, it has contacts on the front as well as the back. But the invention of the prismatic cover means that despite very heavy grid coverage, the classic front-contact design can be used effectively. Adapted to concentration with a heavier grid and a prismatic cover glass, a Green cell has already been measured at 25% efficiency under 200 suns. The cell was 1.2 cm^2 in area, of a practical size for concentrator use. Future directions for improving the Green cell are to make it thinner, reducing recombination, and to make the front grid heavier. As much as 50% front grid coverage can be used if the prismatic cover is used as well. Cells of 30% efficiency under several hundred suns should be possible.

The Swanson cell has received the most attention of the concentrator cells. Developed by Richard Swanson of Stanford University, it has achieved the highest efficiency of any silicon cell: 28.2% under 140 suns. It is an unusual cell in that it uses a thin, highly resistive Si material and polka dots of n- and p-type silicon on the back of the cell to form localized junctions (Figure 35). Light enters the cell and is absorbed in the high-resistivity bulk material. Then the free electrons and free holes migrate to doped regions (electrons to the n-type, holes to the p-type regions). The quality of the silicon is nearly optimal, allowing almost all of the electrons and holes to be collected. All of the cell's contacts are on the back, too—attached to each n- or p-type region—reducing front grid shadowing to zero. This was a solution to the need for a heavily gridded front surface in most concentrator cells,

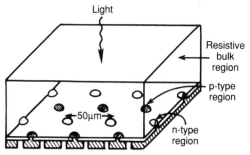

Figure 35. R. Swanson of Stanford developed the point-contact design for silicon concentrator cells. All the contacts and all the n- and p-type regions are on the back. Light is absorbed in a highly resistive, high-quality, single-crystal silicon bulk material. Free carriers diffuse to the numerous junction regions and are separated.

which caused major losses of light to grid shadowing. However, since the invention of the prismatic cover, the need for this solution has been greatly reduced.

High efficiencies are achieved in the Swanson design because there is a minimal amount of recombination. The reason is that junction and contact areas (where recombination can be greatest) are minimized. Meanwhile, the very highest quality silicon is used for the bulk of the cell, also minimizing recombination.

One drawback is that high-quality silicon requires the most expensive silicon growth processes. Another is that the unusual cell design has resulted in an unexpected instability that is very unusual for a crystalline silicon technology. Crystalline silicon is generally regarded as immune to stability issues.

The instability of the Swanson cell has to do with the way the top surface of the cell is passivated. We know that a very thin layer of SiO_2 can passivate bare silicon so that fewer free carriers recombine there. Passivation results from the fact that the oxide ties up most of the dangling bonds at the silicon surface. This type of passivation is the key to many of the gains in the silicon technology, which depends on long diffusion lengths and minimized recombination centers both in the bulk and at surfaces. But under very high concentration of ultraviolet light—the kind encountered at the top of a cell exposed to high concentrations of sunlight—silicon–oxygen bonds tend to break down. The oxide becomes unstable. Fewer dangling bonds are passivated, recombination increases, voltage and current fall. This is the instability observed in the Swanson cell. But work is under way to substitute other passivating layers for SiO_2. That and other strategies should result in the elimination of this surprising effect.

Some effort has gone into commercializing the Swanson cell. Swanson has set up a company to develop manufacturing techniques. In quantity, he estimates that cost will be about $1/cm². At 200 suns (50 cells/m²), cell costs would be about $50/m²—in the range needed for a $200/m² module, and thus appropriate for a 12 cent/kWh$_{AC}$ technology.

The other key single-junction approach is based on gallium arsenide cells. The GaAs technology boasts several very high-efficiency single-junction cells. At least four companies [Varian (Palo Alto, CA), Spire (Bedford, MA), Kopin (Taunton, MA), and Boeing (Seattle, WA)] have made 24%-efficient (or more) *1-sun* GaAs cells. Under concentration and with a prismatic cover glass, a Varian cell achieved the highest efficiency of *any* single-junction cell: 29.2% (206 suns).

The problem with GaAs cells is cost. Each cell is made on a wafer of single-crystal GaAs. These wafers are the result of a crystal growth process like the ones used to grow single-crystal silicon. *A square centimeter of single-crystal GaAs wafer can cost about $1.* Ten thousand square centimeters (1 m²) would cost $10,000—a far cry from the potential of $50/m² or less for thin films grown on low-cost substrates.

GaAs *cells* are typically grown atop single-crystal GaAs wafers by a chemical vapor method that depends on expensive, highly toxic, organic chemicals. The growth must be done on a single crystal because otherwise a non-single-crystal cell (with poor efficiency) would result. After processing, the final GaAs cells may cost about $10/cm², although long-term goals for these same cells are closer to $4/m². At 1000 suns, where 10 cm² of cells is required, cells at $10/m² would cost about $100. This is too much, even given their efficiency potential of over 30% under concentration. But cells costing $40/m² (the long-term goal) would be a big improvement. For single-junction GaAs cells to have an impact, they must be made at these low costs and must approach their theoretical efficiency limits. Even then, they will not be good enough to make 4.5 cent/kWh electricity.

A very interesting approach to reduced cost is being tried at Kopin Corporation. With SERI funding, a team led by John Fan and Ron Gale has found a way—called CLEFT—of growing GaAs cells on *reusable,* single-crystal GaAs substrates. After a thin film of GaAs is grown by a chemical vapor process, the thin film is physically cleaved (viz., CLEFT) from the crystal substrate and processed into a high-efficiency cell. The cleaving process is a straightforward one because a mismatch between the substrate and the GaAs film is purposely incorporated during film deposition. Kopin has reached over 24% with this potentially low-cost approach. The thick, single-crystal GaAs substrate is then cleaned and reused to make another thin cell. In this way, the substrate cost is reduced to a small fraction—depending on the number of reuses—of its original cost. This takes about $1/cm² from the cell cost. If the subsequent chemical vapor processing can be reduced in cost, and production levels can be increased to gain economies of scale, the Kopin approach might yield superior, high-efficiency concentrator cells costing less than $3/cm².

Multijunctions

In some sense, multijunction concentrator cells are the ultimate in PV. As stated above, they have the potential for exotic efficiencies exceeding 45%. In terms of their structure and manufacturing processes, they are extremely complex. Yet very efficient multijunction cells have been made. Only in the last year or so—after a decade of development and more than a decade of speculation—have they surpassed their "prosaic" single-junction cousins. But now they seem to be moving at an increasingly fast pace, advancing the frontier of PV efficiency into previously unheard-of realms.

Multijunctions for concentrators have been made in two different ways. As we discussed in Chapter 4, two cells can be stacked on top of each other so that they have separate external circuits (four-terminal cells) or so that they are connected together and have one external circuit (two-terminal cells).

In some ways the four-terminal approach is simpler. A high-band-gap cell is stacked on top of a low-band-gap cell. Each cell is separately connected to an external circuit. Except for being optically stacked, they are like separate cells. The high-band-gap cell is made transparent to light that has less energy than its band gap. This low-energy light goes through the top cell to be transformed into electricity in the lower cell. We can think of this as simply putting a cell under another one to get more energy out of the combination.

A composite, four-terminal concentrator cell achieved 30.1% efficiency in 1988. This was the first cell to surpass this exalted efficiency. It was made by stacking a Varian GaAs cell on top of a Stanford–Swanson silicon cell. The GaAs top cell provided most of the efficiency (about 27%); the silicon cell received very little sunlight and added only another 3%.

In 1989 Boeing Aerospace made a surprising announcement. They claimed the achievement of a 37% cell under 100 suns concentration (with a prismatic cover from Entech Inc.). This startling efficiency has not been independently confirmed, and uncertainties in the solar simulation probably mean that the true efficiency is closer to 34%. But the Boeing cell does represent a major advance over existing efficiencies. The device was a four-terminal combination of a GaAs cell on top of a new, very low-band-gap (0.7 eV) alloy of gallium with antimony (Sb)—GaSb. The Boeing cell is an indication of the power of the GaAs-based approach: heretofore unheard-of efficiencies are likely to be attained. The Kopin CLEFT cell is another cell that lends itself to the four-terminal approach. Since it is peeled from its substrate, it can easily be stacked on top of another cell. Unlike other GaAs cells, it does not carry an unneeded GaAs wafer—or its cost—along with it. If the Kopin technology could be adapted to the Boeing materials, a reasonably costing ultrahigh-efficiency four-terminal concentrator cell could result.

In contrast to the four-terminal approach, the two-terminal design has some special requirements as well as some potential advantages. In such a design, stacked cells are connected at a common interface. Current flows between them via a low-resistance contact. But because the cells share a common boundary, they must generate equal current if they are to avoid losses. For instance, if the top cell were to produce less current than the bottom cell, it would limit the current of the bottom cell. Equal current must flow through both cells. Equal current can only be achieved when both cells split the solar spectrum almost equally—and then use it equally well. They can do this when they have different band gaps so that the top cell allows as much light through to the bottom cell as it absorbs. Another way to equalize current is to make the top cell thinner than it would be normally (as is done in two-junction a-Si cells made from the same band gaps). Then some energetic light gets through the top cell and can be used in the bottom cell. But overdependence on this method does not lead to optimal efficiencies since it defeats the purpose of multijunctions—using high-energy light in high-band-gap cells and low-energy light in low-band-gap cells.

Because of their performance and flexibility, GaAs-based materials dominate the two-terminal approach. The band gap of gallium arsenide, like that of other semiconductors, can be altered by mixing it with another, suitable element. In its pure state, the band gap of GaAs is 1.4 eV, which is almost perfect for a single-junction cell. However, the optimal two-junction cell in terms of ultimate theoretical efficiency would be made of a 1.75-eV top cell and a 1.0-eV bottom cell. Such a combination would split the solar spectrum into two equal parts, the necessary prerequisite for efficient performance when cells are connected in a two-terminal design. Certain GaAs alloys (e.g., GaAlAs and GaInAs) have these band gaps and can split the spectrum. Their combination has a theoretical efficiency, under concentration, of over 43%.

But perfect band gap combinations are not required to achieve high efficiencies. Well-matched cells in other combinations—0.7 and 1.4 eV—can also split the spectrum equally and potentially reach efficiencies over 40%. This is important because a severe constraint on monolithic cells is that they must be grown one atop the other. When the lattices of the different materials are very different, poor growth ensues. Despite a single-crystal substrate, the material grown on top can be rather poor and make inefficient cells. Thus, technologists must compromise by choosing materials with good compatibility in terms of growth while maintaining reasonable efficiency potentials.

A Varian multijunction cell, based on GaAs alloys and using an Entech prismatic cover, achieved the highest 1-sun efficiency ever, 27.6%. It was far from fully optimized, and has yet to be tested under concentration. With a grid designed for concentration, it would be expected to be about 33% efficient under concentrated sunlight.

Meanwhile, a group led by Jerry Olsen of SERI has made strong progress in GaAs-based two-terminal multijunctions. A cell of 27.3% was made by the SERI group in August 1989, approaching the Varian record for 1 sun. But (unlike the Varian cell) the SERI cell had no prismatic cover glass. With a cover glass, it would be expected to be near 30% efficient *under 1 sun*. This would be the highest efficiency ever under 1 sun. Under concentration, it would be likely to have an efficiency of well over 30%. Arguably, the Olsen cell was the best cell ever made, to date (1990).

These kinds of extreme efficiencies, coming from all quarters, represent the emergence of GaAs-based multijunctions from the shadows of their single-junction counterparts.

Cost of Multijunctions

The GaAs multijunctions share the cost problems of GaAs single junctions in spades. In one sense, making two cells on the same wafer could be viewed as only incrementally more expensive than making just one, and that might eventually be true. But at present, making complex multijunctions adds large costs because yields for successful cells are low. The numerous layers involved—some cells may have 30 of them—individually require superior processing control. Growing single-crystal layers atop layers that are not quite the same—as must be done in alloy cells—poses serious technical barriers. It has been done, but reproducibility is a problem. In addition, the same chemical vapor processing costs that plague the single-junction GaAs technology must be solved in order for the multijunctions to be cost-effective. Even at 35% modules and 1000 suns concentration, we can only afford about $3.50/cm^2 for cells if we are to achieve 4.5 cent/kWh electricity.

Perhaps finding ways to reduce the cost of concentrator cells to their materials costs—which for thin films such as those in CLEFT cells are quite low—will be possible, at least in the long term. Multijunctions can be grown on CLEFT cells, which would eliminate substrate costs. Perhaps yields will increase, or else simpler, four-terminal approaches might be chosen. The payoff might be cost-effective concentrator modules of truly exotic efficiencies, capable of producing electricity at very low costs.

New Ideas

As with all other areas of PV, we have only begun to explore the possible options concerning concentrator systems. Unfortunately, many potentially attractive ideas have received very little attention. The reason (as with all of PV) is the same: With small resources of money and scientists, only a few alternatives are likely to be tried. Some very exotic and potentially exciting areas have never been fully investigated. One of the most interesting ideas is using mirrors (instead of lenses) to focus light. Mirrors can be assembled into very large arrays to focus sunlight. Extremely high concentrations can be reached: over 10,000 suns. As we know, this would change all the economics. Cells could cost an order of magnitude more and still be economical because fewer of them would be used.

Large areas of mirrors (instead of numerous Fresnel lenses) would focus sunlight on a small area of cells—rather than on individual cells spaced a lens apart. The cost of this small array of cells (1 ft^2 of cells for every 1000 m^2 of mirrors) would be almost all cell cost; module costs would be lower than in the lens approach. But one negative result would be that heating effects would be much more severe. All the sunlight would be focused in one small area. How could the heat be dissipated? Active cooling would be necessary. Water or some other medium would need to be circulated. This would introduce mechanical complexities and potential operation and maintenance costs. On the positive side, perhaps the heat carried from the cell array could be used in a thermal converter to increase overall system efficiency.

Another serious problem is that few existing cells work at 10,000 suns because of series resistance and high light-generated carrier densities. Device specialists have rarely approached this problem, so there may be several solutions that are presently unknown or being held proprietary. One characteristic of such a cell would be high voltage and low current to avoid series resistance losses, which would otherwise be 10^8 more than those in 1-sun cells.

This is just one example of the many ideas that have never been fully investigated. We may assume that as the PV business expands and flourishes, these exotic pathways will receive their due.

13 ☀ Sunlight

When we are curious about the amount of sunshine in a particular region, it's normally because we're thinking of moving or visiting there. We may think of Miami as a quintessential sunbelt city and New York as its opposite. Yet measured sunlight in these two cities is very similar during the summer because of the cloudy conditions in Miami and the longer days in the North. From the standpoint of PV, that can be important, because some near-term applications of PV (for summer air-conditioning) depend more on summer sunlight than annual sunlight. We need to know more about sunlight—the solar resource—and its variations, in order to have a more precise idea of how to use PV successfully.

Total Sunlight

Figure 36 shows the total annual sunlight in the United States. The units (MWh/m²) indicate how much sunlight is available locally on a yearly basis. That is the amount that would be available to a flat-plate PV system with a two-axis tracker. For instance, in Kansas City there is about 2.5 MWh/m² of sunlight available each year. The two most important variables controlling the amount of annual sunlight are latitude and local cloud cover. Figure 37 is a map of the average cloud cover over the US. The East and the Northwest are the cloudiest regions. The US desert Southwest stands out in terms of its lack of clouds. The amount of cloud cover in Figure 37 is taken into account in the solar map of Figure 36.

219

How Much Energy Does the Sun Provide?
Average Annual Solar Radiation

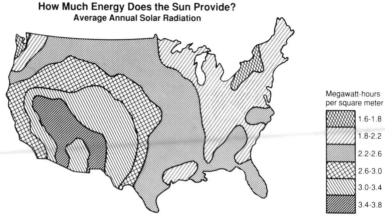

	Megawatt-hours per square meter
	1.6-1.8
	1.8-2.2
	2.2-2.6
	2.6-3.0
	3.0-3.4
	3.4-3.8

Figure 36. The total annual global sunlight available in different locations in the US. This is also the amount available to a flat-plate PV module mounted on a two-axis tracker.

Latitude is important to the amount of annual sunlight for two reasons: the angle of the sun and the length of the days. Although days are longer in winter in the lower latitudes, they are significantly longer in summer (when there is usually more sunlight) in the higher latitudes. This affects the distribution of sunlight seasonally—smoothing the differences between North and South.

The angle of the sun is the final variable affecting solar intensity. As the angle of the sun becomes steeper in the sky, more sunlight is absorbed, scattered, or reflected in the atmosphere. Less reaches the Earth's surface. There is a number that characterizes this effect called the air mass (AM) number. Figure 38 shows how AM is defined. It is actually the secant of the angle of the sun as measured from the zenith (straight up). Put simply, it is the length of atmosphere that light traverses before it reaches the Earth's surface as compared to the length it would traverse at noon on the day the sun was straight overhead. For the sun straight overhead, the AM number is (secant $0 =$) 1. At sunset, the AM is very large: the sunlight traverses the equivalent of many thicknesses of the Earth's atmosphere. On the other extreme, a special number, AM 0, refers to the solar spectrum in outer space, before any atmospheric interactions reduce its energy content.

The solar spectrum changes in two ways as the sun angle (air mass) changes: (1) there is less energy in the spectrum at higher AM numbers (more losses in the atmosphere) and (2) more photons lose

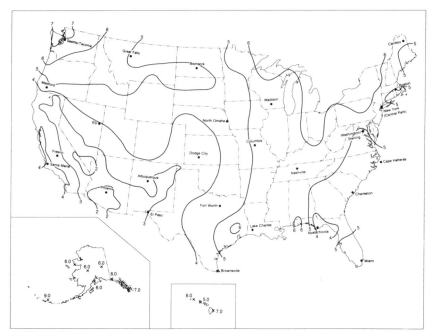

Figure 37. Average daily cloud cover over the US as measured in "tenths of sky," meaning the fraction of sky covered by clouds, as an annual average.

energy through scattering, which has the effect of shifting the spectral distribution toward lower energies. Spectral shifts have a small, but noticeable effect on PV cell performance, depending on the sensitivity of the PV cell to each part of the spectrum. A high-band-gap cell (e.g., a-Si) might lose efficiency in relation to the shifted spectrum, because such a spectrum has a larger proportion of low-energy light to which the a-Si cell is completely unresponsive (the photons that are below the band gap of a-Si). But a cell sensitive to lower energy light, like copper indium diselenide, might gain in efficiency with the shifted light, since it uses the longer wavelength light very well (so it would produce an unchanged output from a spectrum with less total energy).

In practice, variations in AM (Figure 39) are not very critical for most PV systems. This is because, on a daily basis, most sunlight arrives during the hours in which the sun is highest in the sky. Even if a cell becomes less efficient at sunset, system losses are minimal because there is less total energy in the sunlight at those times. The

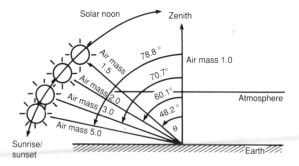

Air mass = Path length of the solar beam through the atmosphere
= Ratio of the path length along the oblique trajectory to the path length
in the zenith (vertical) direction

Figure 38. The air mass (AM) number is defined as the path length of sunlight through the atmosphere. An air mass of 1.5 is taken as the path length for the conventional solar spectrum. The longer the path length, the greater is the attenuation of the light and the less energy for conversion.

only possible exception to this is two-terminal multijunctions used for flat plates—e.g., a-Si multijunctions. In this case, efficiency can be a sensitive function of splitting the spectrum equally between cells. Such multijunction cells are designed to be most efficient (split the spectrum perfectly) at noon on sunny days. At other times of day

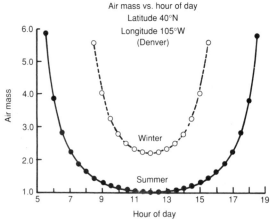

Figure 39. In typical locations like Denver, air mass varies more strongly in the winter than in the summer due to the shorter days and the lower sun angle. Whereas over 12 hours of good sun is available in the summer, only 6 is available in the winter.

(different AM) or under cloudy conditions, such cells are likely to be less efficient because the different spectrum will no longer be exactly split between cells. In this way, two-terminal multijunctions for flat plates are somewhat similar to concentrators—they work best under near-perfect conditions. However, the effect is not as great and not as limiting as it is for concentrators. Again, the problem is most extreme only when the spectrum is very different from the noontime spectrum, i.e., when it is cloudy. In terms of energy production, these times of poor sunshine have the least overall impact, anyway.

The Conventional Spectrum

Although the AM number can be used as a parameter to characterize the solar spectra found at different solar angles, it does not uniquely define a spectrum. The amount of water vapor, turbulence, haze, or cloud cover can affect sunlight significantly as well. They are not part of the AM definition. To facilitate measurements and the intercomparison of PV technologies, a conventional solar spectrum has been established. This spectrum was designed to be typical in terms of average values of most phenomena (e.g., turbulence, water vapor), excluding cloud cover, which is hard to characterize. The conventional spectrum is a clear sky spectrum. Figure 40 shows the distribution of photons in the conventional spectrum. AM 1.5 (not AM 1) has been chosen as the sun angle in the conventional spectrum, because this angle is more typical of PV operations than perfect noontime conditions would be. The standard AM 1.5 spectrum of Figure 40 is used in the efficiency measurements of all PV devices.

The spectrum of Figure 40 shows the number of photons arriving at the Earth's surface as a function of their energy content (in electron volts) or wavelength (in microns). Large gaps appear in the spectrum at lower energies. These troughs occur at energies where constituents of the atmosphere absorb sunlight before it reaches the Earth's surface. In the range of interest for PV, most losses are caused by light being absorbed by molecules of water vapor.

On the other end of the spectrum—at higher energies than 3.0 eV (ultraviolet light)—almost all sunlight is absorbed by atmospheric ozone high in the atmosphere. From the standpoint of human health, this is very important. Even in small doses, high-energy, ultraviolet

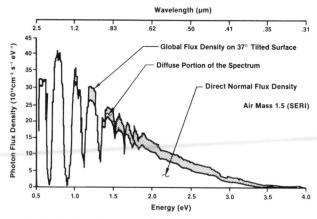

Figure 40. The distribution of photons in the conventional SERI global AM 1.5 spectrum and a comparison of that distribution for both global and direct sunlight. The direct spectrum has fewer photons at higher energies because these are the photons that have been scattered during their passage through the atmosphere.

light can cause skin cancer. That is why environmentalists have been so disconcerted by observations that the ozone layer is being depleted by chemical reactions with man-made chlorofluorocarbons (CFCs). PV devices would produce marginally more power if the ozone layer were fully depleted, but mankind would have to wear #15 sunblock all the time or live underground.

At very low energies (below 0.5 eV), significant absorption of *infrared* sunlight is caused by water vapor, carbon dioxide, methane, CFCs, and other greenhouse gases. These gases are the source of the greenhouse effect. Not much energy comes to Earth from the sun in these longer wavelengths. But most of what the Earth reradiates back to space travels in the form of infrared radiation. This reradiating infrared can be blocked—kept in the atmosphere—by the greenhouse gases. We are most familiar with the greenhouse effect in terms of water vapor. Most of us know that summer nights stay warmer when they are humid. Places like the Southeast have small variations in their daytime highs and nighttime lows. In contrast, desert evenings can be quite cool because the atmosphere has less water vapor in it, and infrared radiation can effectively escape to space.

Atmospheric scientists tell us that without the greenhouse effect caused by naturally occurring water vapor and carbon dioxide, our planet would be much cooler—about 55° F cooler (which would make the Earth almost unlivable). Scientists recognize the greenhouse effect as one of the fundamental phenomena controlling our global climate. That is why they have become concerned about the growing greenhouse impact of carbon dioxide and other trace gases (e.g., CFCs). Amounts of carbon dioxide and other greenhouse gases have been increasing rapidly during the last two centuries. Carbon dioxide was about 270 parts per million (ppm) in the preindustrial era. Now (Figure 41) it has reached about 350 ppm and is increasing at over 1 ppm annually. Consciousness of the potential hazards emanating from the greenhouse effect—in large part resulting from the combustion of carbon from fossil fuels—is perhaps the major driving force behind the rebirth of general interest in PV.

Figure 40 shows another important aspect of the spectrum: it has two components, a direct component (under the lower curve) and a diffuse component (the stippled portion). We have frequently noted that some PV devices (e.g., concentrators) cannot use the diffuse portion of the spectrum. Together, the direct and diffuse spectra make up the total, or global spectrum. The direct component is the sunlight

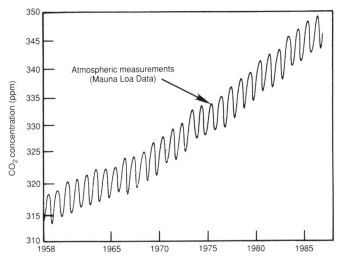

Figure 41. The concentration of carbon dioxide in the atmosphere has increased from about 270 parts per million (ppm) before the industrial age (pre-1700) to about 350 ppm today. Carbon dioxide is emitted by all fossil-fuel combustion processes, with the largest proportion coming from energy production like coal burning.

we see as coming from the sun. The diffuse portion is the rest—sunlight that is refracted and scattered in the atmosphere. It appears to us as blue sky on a cloudless day and hazy sky light on a cloudy day. On a cloudless day, about 10–20% of the light is diffuse. Under cloudy conditions, almost all of the light is diffuse (over 90%). The more turbulence or haze is present in the atmosphere, the more diffuse light. Desert locations have the least diffuse light; a hazy area like New York has much more, even on a cloudless day. In northern Europe, about 50% of the annual sunlight is diffuse.

Trackers and Sunlight

The fact that sunlight has diffuse and direct components bears rather significantly on the economics of PV. It determines (1) the value of trackers and (2) trade-offs between flat plates and concentrators. Figure 42 characterizes the ways in which trackers can take advantage of both direct and diffuse light. Concentrators (left) follow the sun across the sky but cannot focus diffuse light; they only use direct sunlight. Tracking flat-plate arrays (middle) follow the sun and use the total, global spectrum. Fixed flat-plate arrays (right) use the global spectrum, but lose a portion of the direct sunlight because it comes in at an angle as the sun moves across the sky.

The angular losses that fixed flat plates experience have two parts. One is from reflection losses, which increase with the increasing

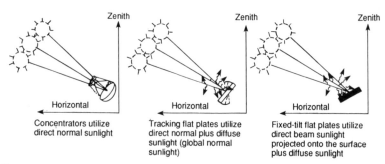

Figure 42. Concentrators (left) track the sun but use only direct sunlight; flat plates on two-axis trackers (center) use the total global spectrum; fixed flat plates receive global sunlight at an angle that varies with the position of the sun (right).

oblique angle of the sunlight. The other loss is from the fact that actual PV collection area becomes smaller as the sun moves away from being perpendicular to the array. We can visualize this by considering how much sunlight a sheet of paper would intercept as we turned it away from the sun. The paper would intercept less and less sunlight until at the extreme, it would intercept no sunlight except at its edge.

There are many possible ways to track the sun, but four are the most important: fixed flat plates, 1-axis trackers, 2-axis trackers, and concentrators. Fixed flat plates are usually set up pointing south (in the Northern Hemisphere) at an angle from the perpendicular (straight up) equal to the local latitude. This is an average position that is designed to minimize losses both seasonally and daily. In some cases, the angle of the array is not the latitude angle but is chosen to maximize solar availability (and system output) during a chosen time of day or season. For instance, the array may be aligned a little west of south, if late afternoon output is desired, because that is when the customer has a peak requirement.

Single-axis trackers are tilted at some compromise angle (usually the latitude, again) but then mechanically rotated about one axis so that they follow the sun during the day. They can very accurately minimize angular mismatches on a daily basis, but they cannot change their angle in relation to the southern horizon as the track of the sun moves up and down with the seasons. One-axis trackers intercept more sunlight (about 10–20%) than fixed arrays but require more sophisticated structures that are significantly more expensive.

Two-axis trackers are the most complex and sophisticated trackers, and they can be used for either flat plates or concentrators. They track the sun across the sky so that it is always very nearly perpendicular to the array. This maximizes solar availability. For flat plates, they might be considered a luxury, since they intercept very little more than single-axis trackers (less than 10%) and can be quite a bit more expensive. For concentrators, they are a necessity, since concentrators require that the sun be perpendicular to lenses that focus sunlight on cells. In fact, two-axis trackers designed for concentrators are the Cadillacs of their kind, since they are made for very precise solar tracking. As such, they are the most expensive and most likely to have maintenance problems, especially in windy or dusty areas.

Unfortunately, the relationship between trackers and sunlight is complex. No simple formula can be used to relate the total sunlight available for a two-axis tracker (Figure 36) with sunlight available to other trackers or to a fixed array. That is a function of the specific geometry of the tracker and of the diffuse and direct components of the local sunlight. But in general:

☐ Fixed flat plates can use both diffuse and direct sunlight, but some portion of the direct sunlight is lost because of oblique sun-angles in relation to the array.

☐ Single-axis flat plates use both direct and diffuse sunlight, and they use the direct better than the fixed arrays (but not perfectly).

☐ Two-axis trackers for flat plates use the total global spectrum almost perfectly (Figure 36).

☐ Two-axis concentrator trackers use the direct spectrum perfectly, but they get nothing from the diffuse portion, which cannot be focused on their cells.

A series of figures (36, 43, and 44) can show the different amounts of energy available to different kinds of trackers. Figure 36 shows the total, global sunlight. That would be the amount of sunlight available to flat plates mounted on two-axis trackers. It is the largest amount of available sunlight for any kind of PV system. However, a two-axis tracker may not be economical from a total system cost/output perspective unless tracker costs drop faster than expected. A different rationale for two-axis tracking is that it extends the period of PV electricity production because the tracker can follow the sun early in the morning and late in the afternoon. This may be of use in various applications that require more electricity at these times. For instance, the peak need for electricity is often in the late afternoon. Trackers would help PV to meet this need.

Figure 43 shows the amount of *direct* sunlight available for two-axis trackers. This is the amount that a concentrator would use. Note that it is always less than the amount in Figure 36, since it is without the diffuse portion. As mentioned, the difference between global and direct sunlight is generally least in the most arid regions like the Southwest (see also Figure 37, cloud cover), because the greatest cause of a large diffuse component is cloudiness. But in regions like the East Coast and the Northwest, direct sunlight (Figure 43) is significantly lower than global sunlight (Figure 36). Direct sunlight can be

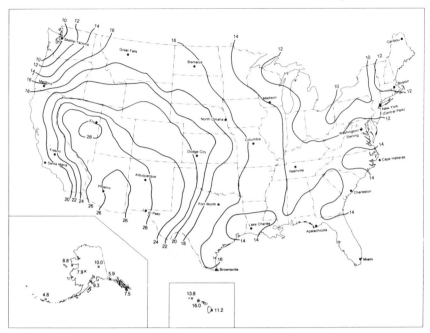

Figure 43. Average annual direct sunlight (in 100s of kWh/m²) available to a two-axis tracker. This is the amount that is usable by a concentrator module.

so reduced that it can actually be less than the amount of sunlight a fixed flat plate—with all its losses—would receive in the same location. That effect is shown in Figure 44, which shows how much sunlight would be available to a fixed, flat-plate array. These amounts are significantly below those for two-axis flat plates. In arid areas, they are below those for concentrators (two-axis direct; Figure 43). But in most regions, they are very similar or even higher than sunlight for concentrators. This illustrates the problem of locating concentrators in non-arid regions: They might very well get too little sunlight to offset their added cost.

Table 13-1 shows the problem of concentrator location in more detail. For the Southeast US, it shows the annual amount of sunlight available to (1) a fixed array, (2) a single-axis array, and (3) a two-axis array, as a ratio of that which would be available to a two-axis concentrator (which, as we know, can use only direct sunlight). Note the fact that nowhere in the Southeast does the concentrator ever receive even

Table 13-1. Comparison of the Seasonal Solar Availability for
Concentrators versus Flat-Plate Collectors in the Southeast[a]

	Annual ratios		
	Conc/FFP	Conc/SXTFP	Conc/DXTFP
Atlanta	.826	.71	.636
Austin	.862	.725	.66
Birmingham	.806	.694	.625
Miami	.763	.654	.6
Nashville	.775	.68	.61
New Orleans	.806	.69	.63
Oklahoma City	.893	.763	.676
Orlando	.77	.66	.6
Raleigh–Durham	.735	.65	.58
Average	.8	.69	.62
Las Vegas	1.06	.88	.77

[a]FFP: fixed-tilt (latitude), south-facing flat plate insolation, average daily. Conc: two-axis tracking concentrator. SXTFP: single-axis (north–south, horizontal) tracking flat plate. DXTFP: two-axis tracking flat plate.

as much sun as a simple fixed array would. The value for Las Vegas (1.06) is included to show that this is not universally true: in the Southwest, the diffuse component is much smaller, which means that there a two-axis tracker of direct sunlight can get more sunlight than a fixed array using global sunlight.

Table 13-1 also shows the variation among different trackers, with single-axis getting about 10–20% more sunlight than fixed; and dual axis flat plates getting another 10% over single-axis. These improvements are fairly typical, even in other regions.

Table 13-2 shows the amount of sunlight in various US cities during different seasons. The table is useful for decisions based on seasonal requirements, e.g., the need to maximize output during the summer. Also note some of the peculiarities of these numbers (from Table 13-2):

□ In the summer, Miami receives less sunlight than in the spring, and about the same amount as in Caribou, Maine (or Washington, DC).

□ Denver—generally regarded as a snowbird capital—gets more sunlight in the *winter* than does Miami, Honolulu, New Orleans, Sacramento, or San Juan. On an annual basis, it gets *far* more sunlight than these other cities.

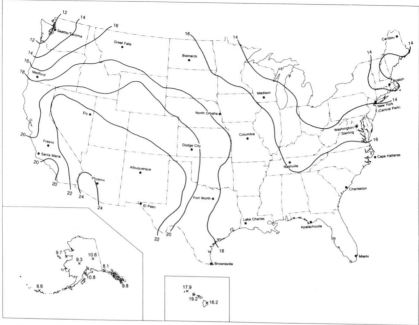

Figure 44. The average annual global solar radiation (in 100s of kWh/m²) on a south-facing surface tilted at an angle equal to its latitude. This is the amount of sunlight available to a fixed flat plate. (Compare with Figure 36, the global sunlight available to a two-axis tracker.)

□ Fairbanks, Alaska, gets about the same amount of sunlight in the spring and summer as do Miami, Nashville, New Orleans, and Washington, DC. Northern latitudes can use PV very well during their long summer days.

Conclusions

Some people think that PV can only be used in sunny and warm places like Denver or Phoenix. But for flat-plate PV almost 90% of the US has sunlight that varies on an annual basis by under 50% (see any of the figures, especially Figure 44; global on fixed flat plate). Only the most extreme desert regions and worst rainy ones fall outside these limits. For instance, the annual total sunlight in New York is about 2 MWh/m²; in Denver, 3; in Florida, about 2.4; and in California,

Table 13-2. Average Daily Sunlight Available
for a Full Tracking Surface (kWh/m²)

	Winter	Spring	Summer	Fall	Annual
Albuquerque	7.43	10.41	10.87	9.04	9.45
Atlanta	4.44	7.15	7.53	5.91	6.27
Austin	5.53	6.88	8.41	6.58	6.86
Birmingham	4.33	6.89	7.10	5.97	6.08
Bismarck	4.09	7.16	9.34	5.47	6.53
Boston	3.23	5.82	7.14	4.78	5.25
Brownsville	4.92	7.20	8.85	6.35	6.84
Bryce Canyon	7.22	10.50	11.01	8.85	9.41
Caribou	3.33	6.73	7.25	3.85	5.30
Columbia	3.98	6.97	8.71	5.92	6.41
Daggett	6.68	10.36	11.50	8.47	9.27
Dallas–Fort Worth	5.08	7.09	9.10	6.81	7.03
Denver	6.70	9.27	9.90	8.04	8.49
Detroit	2.89	5.97	7.47	4.57	5.24
Elko	5.90	9.24	11.66	8.12	8.75
El Paso	7.76	10.65	10.72	8.96	9.53
Fairbanks	0.66	6.79	6.97	2.44	4.24
Fresno	4.32	9.75	11.79	8.09	8.51
Great Falls	3.68	7.20	9.87	5.79	6.65
Honolulu	5.99	7.25	8.04	6.95	7.06
Las Vegas	7.37	11.22	11.40	8.86	9.73
Madison	3.62	6.62	7.79	4.90	5.75
Medford	2.57	7.22	10.55	5.58	6.50
Miami	6.00	7.35	6.55	6.19	6.53
Nashville	3.61	6.26	7.64	5.41	5.74
New Orleans	4.79	7.15	7.20	6.29	6.36
Oklahoma City	5.31	7.49	8.77	6.79	7.10
Omaha	4.90	7.29	9.00	5.98	6.81
Orlando	5.73	7.80	6.91	6.27	6.69
Phoenix	7.07	10.76	10.59	8.75	9.30
Pittsburgh	2.52	5.46	6.89	4.36	4.82
Raleigh–Durham	4.06	6.75	7.26	5.38	5.88
Sacramento	4.20	9.13	11.68	7.59	8.18
San Diego	6.23	7.99	8.52	6.99	7.44
San Juan	6.61	7.30	7.52	6.77	7.05
Seattle	1.76	5.68	7.93	3.66	4.78
Syracuse	2.26	5.49	6.98	3.82	4.65
Washington, D.C.	3.90	6.45	7.30	5.12	5.71

between 2.6 and 3.0. The upshot of this is that PV is not geographically limited in the US. This is very important to recognize for the future of PV in America.

Besides the myth of geographical limitation, another myth concerning the inapplicability of PV has been that sunlight does not provide enough energy. This is usually summed up in the statement "sunlight is a diffuse form of energy," implying that significant energy needs—those of our affluent life-styles—could not be met by PV. The opposite is closer to the truth: PV is one of the easiest ways to meet our growing energy needs. A conservative estimate of the annual sunlight falling on the US is found by multiplying the nation's area by an average amount of sunlight: (6.4×10^{13} m$^2 \times 2000$ kWh/m$^2 = $) about 10^{17} kWh/yr. The US now uses about 2.5×10^{12} kWh of electricity per year. This is equivalent to the energy of about 0.002% of the sunlight falling on the US. We use 2.4×10^{13} kWh of total energy (including electricity), a much larger figure because it includes primary energy used for electrical generation as well as fuels for the transportation sector. This is about 0.02% of the sunlight on the US.

Knowledge of the solar spectrum allows us to locate PV properly and to design support or tracking structures for local needs. It also provides critical insights that to some extent determine how we design our PV devices. At the most basic level, however, it tells us that PV can tap into a vast and constantly renewed source of energy, sunlight, and that such energy is sufficient and widely enough available to be of practical use.

14 ☼ Today's Market for PV

PV is now being used for a diverse set of applications, but only on a very small scale. The total world sales of PV is about $400M annually, or about 40 MW of PV systems. This is a drop in the bucket as far as world energy is concerned, but the present market for PV provides a valuable foundation for the future:

☐ Existing markets provide companies with partial displacement of R&D costs, since they can use sales revenue to fund R&D.
☐ Sale of products provide a near-term basis for investment in PV companies.
☐ PV applications are a laboratory for developing new products and expertise that will be of significance for future, larger applications.
☐ They stimulate public awareness of PV.

Present PV systems fall roughly into the following categories:

☐ Remote power
☐ Consumer electronics products
☐ Demonstrations

Remote Power

People remote from utility electric grids have to generate their own power. This applies to some rural locations in the US and to most places in the Third World. The remote power market has always been a key market for PV because it represents, for the most part, truly competitive applications. Also, the degree of remoteness at which PV becomes competitive is a good measure of the maturity of PV as a technology, since as PV becomes less expensive the requirement for remoteness will diminish. When PV becomes truly competitive, it will have moved from the most remote market of all—space—to our own backyards (or roofs).

In existing markets that are isolated from conventional utility electricity, some unique aspects of PV (such as independence from ongoing fuel requirements, little need for maintenance, and modularity) can be much more significant than the cost of PV. Conventional methods for making electricity in isolated locations (e.g., diesel generators or extending the utility grid) may not be feasible. Or they may not be useful because of the need for constant refueling and maintenance. In some places (e.g., the Fiji Islands), these considerations make all the difference. Natives cannot be expected to successfully maintain a generator that requires frequent, technically sophisticated attention.

The most obvious example of PV's competitiveness in isolated markets is its original one: outer space. No other method of generating electricity has been able to match its convenience, reliability, and flexibility in outer space. The need by NASA and the Air Force for PV devices has provided a valued market for PV for 30 years. Many of the advances associated with very high-efficiency cells can be traced to this ongoing interest.

In this chapter, we will discuss other, terrestrial markets, that, like the outer space market, are isolated from conventional electricity sources. These markets for PV include the use of PV in remote locations for water pumping, vaccine refrigeration, rural electrification, desalination, cathodic protection, communication, warning lights and beacons, lighting for signs and billboards, lighting for bus shelters, and monitoring (e.g., weather data). Each of these small uses takes advantage of PV's unusual capabilities to meet needs that would be very hard to meet using conventional energy sources.

Water Pumping

Using PV to pump water in isolated areas has been an important PV application. About 20,000 of such systems are operating worldwide. The water is usually needed for drinking or irrigation. In the US, PV water-pumping is also used for pond aeration (keeping enough oxygen in the water for the fish) or at campgrounds for drinking water. In isolated places, the common means of getting water (without PV) is manual pumping, performed by human or animal power. Another way is to use diesel-powered generators, which can be very expensive and inconvenient. Even at its present cost, PV is competitive with almost any conventional method of pumping water in an isolated location.

One of the most convenient aspects of a PV water-pumping system is that water storage can easily be incorporated. When the sun is out but the water is not being used, water can be continuously pumped by PV and stored in an aboveground storage tank. When it is needed, the water can be drained from the tank and used, even at night. Thus, no PV power is wasted; and water is always available. Using this system solves PV's classic problem of intermittency. PV is also an exceptionally good match for irrigation, since the ground acts as a storage medium, and since more water is usually needed when it is sunnier.

The size of PV water-pumping systems ranges from a few hundred watts to a few kilowatts. The classic use of PV for water-pumping is in Third World communities, where it replaces human or animal power. The Indian government has distributed about 900 PV pumps. A typical system may be as small as 0.4 kW and consist of a PV array, support structure, wiring, and a switch and motor coupled to a DC pump. Such a system is capable of lifting 20,000 liters of water 7–8 meters. The Indian government has found PV preferable to the use of diesel power because PV does not require refueling and needs little attention for operation or maintenance. Low-tech maintenance is very important in all of these applications.

In the US, water-pumping systems have been used to provide water in campgrounds or in remote locations for irrigation or to water cattle. An example of the latter is a system in Owyee County, Idaho, capable of providing 7500 gallons of water a day. The system uses a submersible pump powered by a 1.1-kW array mounted on a passively driven, one-axis tracker. The system has had an excellent operational record since its installation in March 1987. It cost $12,000 ($11/W$_p$) and provides water to livestock grazing on open range managed by the Bureau of Land Management.

An unusual application of PV involves the aeration of a pond at the Walter Heller Nature Center, Highland Park District in Illinois. The pond's water was of poor quality for the fish because of a low dissolved-oxygen content. Now, a PV system pumps air through the pond during the day. The system cost $3500 to install and provides a unique value that would be hard to reproduce by any other means suitable for the park environment.

The cost of PV is presently too high to be fully cost-competitive in the US for rural (off-grid) irrigation. However, with PV system costs dropping to $2–$3/W_p in the late 1990s, this use could grow significantly. Present systems use 100-kW electric motors to pump water. They usually require line extension from the local rural utility. They have significant costs associated with electricity, line extension and maintenance, and annual hookup charges. Another trend may also favor PV: the use of underground watering systems to conserve water. These would require less power than conventional irrigation facilities, so smaller PV systems would be able to meet their energy needs. PV could become a popular source of energy for new irrigation in the US during the late 1990s.

Village Power

Another classic application for PV is electricity for remote or Third World villages. PV electricity can be used for a variety of things—emergency medical power, vaccine refrigeration, lighting, radios, televisions, and telephones. Systems are fairly small: 1–25 kW_p. In this context, very efficient, low-power appliances are often used in conjunction with PV, since they allow for more diverse needs to be served. Similarly, DC equipment is preferred since DC obviates the need for inverters. In most cases, batteries are also used, to provide storage and power on demand. But as more appliances become part of the village grid, demand can exceed available power. To avoid problems, demand-side controls are usually built into these systems. Such computer controls can automatically switch off secondary uses like sewing machines, washing machines, and televisions when primary needs such as refrigeration are not being met.

Usually there is no local electric utility grid or even the possibility of a line extension from such a grid. The competition is diesel-

powered electric generators, which have serious operation and maintenance problems. PV is now considered cost-effective in remote areas
in comparison to diesels, when costs over the life of the system are
taken into account. However, the up-front cost of PV is still daunting
in most instances, which has slowed the spread of these systems despite their cost-effectiveness and proven reliability. Complicating the
cost issue is that the need is usually among those with little capability
of paying for it. In this case, government or international aid agencies
take on the financial role of providing capital. It has taken a long time
for various aid agencies to accept such a radical new technology as PV,
which is one of the main reasons that the PV industry has taken so long
to develop. The growth of remote systems—after many years of successful demonstrations and information dissemination—is in large
part fueling the growth of PV today.

The potential market for village power is huge: the Indian government estimates that there are 300,000 villages in that country with
no electricity. Off the coast of Korea, there are 3000 islands currently
using diesel generators for their electricity. Worldwide there may be as
many as 5,000,000 villages that could use PV.

Electricity from PV is highly prized in these circumstances. The
locals usually have no other access to electricity. Its introduction is
like a welcome to the modern world. It allows lighting, communication by telephone, and exposure to television. These are not necessi-
ties, but they are amenities that people will characteristically pay a lot
for and consider extremely important. Under the circumstances, the
beneficiaries of these small PV arrays are very appreciative of PV and
do not even notice the problem of having to curtail various electrical
uses at times of peak demand. To them, the enriching value of PV is
more than can be measured in dollars and cents.

For instance, it is often found that PV electricity (with battery
storage) can be used to create night schools. Such schools are impossible without electricity for lighting. At night (after work), schools can
attract more students. And they may have new, and valuable tools,
such as instructional television, impossible without electric power.

Well-lit community centers with PV televisions provide a central
meeting place for people to socialize. Birth rates can actually drop
when such nighttime alternatives exist. These small luxuries mean a
lot when there are few other choices. PV is uniquely capable of meeting needs in the Third World, especially for people who are unused to

the convenience of our utility grid system and incapable of affording its cost. When PV costs drop to their expected lows in the new century, these locales will be well-positioned to use PV with storage as the basis of their electricity use.

Communication

Another natural use of PV is for powering systems that provide communications (phones, television, radio) in remote locations. Many such applications are for transmitters on hilltops or mountains, where PV with battery storage has clear advantages over systems that require fueling or maintenance.

In the US, there are several thousand PV-powered communication systems:

☐ Secondary transmitters for TV and radios
☐ Microwave repeaters for telephones
☐ Emergency telephone call boxes

System sizes vary from a few watts for telephone call boxes to several kilowatts for microwave repeaters. Applications are always off-the-grid, usually in remote areas susceptible to severe weather, including high winds and heavy snows.

Microwave repeaters are among the most successful existing PV uses. They have battery storage to provide continuous power. The repeaters use microwaves to transmit telephone calls from horizon to horizon in regions where stringing telephone lines would be prohibitively costly. In the US, they are being used in remote areas such as Alaska and the Rocky Mountains. Internationally, they form one of the largest markets for PV. Australia Telecom uses them widely in the Australian outback, where few other systems would be feasible for bringing telephones to this vast, lightly populated region. The alternative would be to power the repeaters with diesel generators, where the expense would not be so much for the fuel as for the transport of the fuel by helicopter to isolated areas where no roads exist.

Emergency telephone call boxes powered by PV (with battery backup) are growing in popularity. In California, the Orange County Transportation Commission installed about 1100 such call boxes along its freeways. These popular call boxes are directly connected to the Highway Patrol and use cellular phones.

Warning Signals

Systems to provide warning signals in isolated areas are another natural use of PV (with batteries) and one of the largest existing PV markets. In the US, such systems are owned by the Armed Forces, the Coast Guard, the oil industry, railroads, highway departments, and many others. All of them have battery storage to allow for continuous availability. Examples of such systems are:

□ Navigation beacons
□ Sirens for floods and other emergencies
□ Highway warning signs
□ Railway signals
□ Aircraft warning beacons

The Coast Guard is using about 10,000 PV-powered systems of this sort, mostly marine navigation aids. These include lights, radios and sirens for lighthouses, and lights for channel beacons. They estimate that they have saved about $5 million a year on these systems versus their previous systems, which consisted of batteries alone and required more maintenance and replacement.

Isolated warning sirens are a growing application. Six such PV-powered systems are in place in Pedernales Falls State Park near Austin, Texas. Each system is mounted on a tower and has a small (20 watt) PV module connected to a battery and a siren. The sirens can be remotely activated in case of severe weather. They warn of the danger of flash flood.

Remote Monitoring

Almost 20,000 small PV systems, most under 200 W, are being used in remote locations in the US for monitoring various phenomena. Examples of such uses (which all have battery backup) are for:

□ Weather and climatic information

☐ Pollution data
☐ Highway conditions
☐ Seismic records
☐ Other research data

Climatic conditions are monitored for the National Oceanic and Atmospheric Administration (NOAA) by numerous PV-powered monitors, some of which have been in place since 1978. These monitors measure temperature, wind speed, and other data, and transmit the information directly to weather satellites. With the help of PV, NOAA is now remodeling some of its 5000 sites used to measure rainfall and the height of water in rivers. Another agency, the Forest Service, has about 500 PV-powered weather monitors in the Rockies and Alaska.

An application of growing popularity is remote pollution monitoring. PV has been used to sample air quality in the Grand Canyon for research purposes. Measuring carbon dioxide and other greenhouse gases in such isolated regions—where data gathering should reflect average global atmospheric conditions rather than local perturbations—is also a growing need.

Lighting

Various isolated lighting needs can be met by PV. In the Third World, PV-powered lights located on the main street of a town can create meeting places for people to gather and socialize. In the US, PV can be used to power security lighting, bus shelter lighting, highway signs, and billboards. About 6000 such systems have been installed in the US. Recent improvements in the efficiency of lights have made use of PV easier for these applications. Lighting for bus shelters is a typical, growing use. The cost of connecting new shelters to existing power lines, even though they are usually nearby, is greatly increased by the need to scale down the voltage of electricity from the grid with a transformer. A large number of PV/battery systems can be installed for the cost of one such transformer (about $3000). Many PV bus shelters are being installed in California and Florida. Another popular use is to light remote areas where security is an issue. Companies can illuminate parking lots or the periphery of a property at great savings over extending power lines.

Other Remote Uses of Power Modules

Numerous other markets are also growing in the US and internationally, including:

□ Off-grid residential power
□ Cathodic rust protection
□ Restrooms and other isolated needs in national parks

Over 17,000 remote US residences, vacation cabins, park visitor centers, and research buildings are being supplied with electricity from PV. These systems usually include battery or diesel backup to drive small, efficient appliances and communications equipment.

Cathodic rust protection by PV is accomplished when a small electrical current from a PV array blocks the natural rust reaction between water and metals. Over a thousand such systems have been installed to protect bridges, pipelines, buildings, wharves, docks, and marinas.

Isolated public restrooms powered by PV are an ideal small-scale application. They can be used in national parks, along highways, and at beaches. Such restrooms usually have exhaust fans, lights, running water, and a toilet that recycles waste—all of which can be assisted by PV electricity. In many cases, PV-powered restrooms provide unique advantages, since the usual alternatives—utility grid extension, diesel fuel, and high maintenance costs—may be impossible.

The use of PV for remote power is an important and growing part of the PV market. It represents applications that are cost-effective. The future of these markets is bright because their economics will improve as PV becomes less expensive. Many years of corporate effort have gone into developing the remote power market. Potential consumers who had need for PV were hard to contact—often in underdeveloped countries—had little money, and knew next to nothing about PV. Funding agencies were often unresponsive to a new, untried method of producing electricity. But successful demonstrations have been done, and the market is improving rapidly. Years of investment and effort are beginning to pay off. Without other intervention (e.g., government stimulus), it is this market upon which much of the hope for PV development rests.

Consumer Products

Consumer products powered by PV were an innovation developed mostly by Japanese and US companies that were developing a-Si PV. These companies were able to take advantage of the low cost of small, a-Si cells and modules to start a market in calculators, watches, radios, and toys. A few electronic products that have been traditionally powered by batteries are becoming available as convenient, PV-powered products. In addition, new products such as small outdoor lights are being developed for yards and patios, and for residential address signs that light up at night. PV is also being used to charge batteries for recreational vehicles (RVs) and to continuously charge car batteries during extended periods of parking. Another new product is the car fan. Powered by PV, the fan exhausts a car of hot air while it is parked outdoors. Other applications by the auto industry, such as PV-powered in-car sensors and computers (without depleting the car battery), are also expected to grow.

The consumer market was originally dominated by Japanese companies. Today, US a-Si companies control much of the market, but newcomers are rapidly joining them. Other materials—crystalline silicon and new thin films—are also being adopted for these products. The future of PV consumer products depends on how well specific items capture the imagination of consumers. Should a particular consumer product catch fire with the public, it could become a major source of revenue for the PV industry.

In their own small ways, all PV products—from village power to calculator cells—contribute valuable knowledge that will lead to the ultimate goal of using PV for large-scale electricity. Although PV is not yet cost-effective for large applications, numerous demonstrations are under way to prepare PV for its future market.

Demonstrations

Demonstrations of PV for large-scale uses fall into three categories:

□ Utility PV
□ PV for residences and businesses

□ Other specific demonstrations

These demonstrations have almost always been funded by government, or by tax credits provided by the government. In the past, they were often done with the purpose of showing that PV systems could be used "today." They did not accomplish their purpose, since the systems were consistently too expensive. During periods of enthusiasm fueled by unrealistic hopes, such as the late 1970s, the PV systems were touted as prototypes of commercial systems. They failed miserably in this role and gave PV a bad name in the process. The fault was not with PV, but stemmed from the short planning horizon that most of us share. When we are ready for something new, we want it immediately.

However, the same systems did show that PV could perform reliably and predictably, and could be integrated into various conventional energy systems, including utilities. Viewed in this context (rather than as prototypes of near-term commercial systems), the demonstrations could be deemed a success. Those that were built on a small scale—as experiments rather than commercial prototypes— were worthwhile. When truly cost-effective PV is achieved, the way will be a bit easier because of these early systems.

The other, larger systems of the late 1970s were major drains on federal money. Instead, the money should have been used to accelerate R&D progress in new, more promising technologies. For instance, funding of innovative thin films like a-Si, CIS, and CdTe did not start until the era of demonstrations petered out in the early 1980s. Even today we may wistfully compare the three or four million dollars we spend annually on CIS R&D with the hundred million spent in 1980 on buying PV systems that cost $50,000,000 for a megawatt.

This problem—focus on immediate commercial systems, rather than long-term R&D—is one of the natural failings of most developmental programs—in government or out of it. We may expect the same problem again in the 1990s as PV becomes better known as a key new source of energy. People will again want PV "today." This, new round of PV systems will be very nearly cost-competitive. Arguably, the situation has changed enough that buying PV systems rather than PV R&D will make sense. But as existing systems are implemented, potentially even cheaper PV technologies may not receive their due proportion of resources.

Central Station PV for Utilities

Many people visualize the ultimate use of PV as systems distributed on rooftops. This sparks their interest because it conjures up images of energy independence from the utility. They may be right about PV. But central station PV—large PV "farms"—could have a general impact as well. In some cases, costs can be lower for centrally located PV. In relation to rooftop arrays, central stations have reduced costs for marketing, transportation, and installation. Instead of marketing to a hundred consumers, with all the associated costs, centralized PV has only one customer, the local utility. Central station PV would also have added flexibility for tracking or use of concentrators. On the other hand, siting could be easier and land costs could be lower for rooftop PV (a free ride atop a building), and some transmission losses could be avoided because the PV would be located where it was needed. PV applications will turn out to be a mix of centralized and distributed alternatives.

The US DOE funded nine utility-connected PV systems between 1980 and 1983. They ranged in size from 17 to 200 kW. Since PV costs were very high at the time (as much as $100/W_p$!), such systems were extremely expensive, in some cases absorbing more money by themselves than the entire PV research program. The money was diverted from research that could have been done in the mainstay of the program at the time, crystalline silicon. It also deprived the research program of almost all funding for advanced technologies such as thin films or high-efficiency gallium arsenide. These latter options have survived and even flourished in the 1980s on funds that have been much smaller than those of the late 1970s and early 1980s—less than $100 million spent during the entire decade as compared to almost a billion prior to that. As is often pointed out by skeptics of PV, it is true that more than a billion dollars has been spent on PV by the federal government. But at least half of that has gone for premature demonstrations of noncompetitively priced systems. Once the government got out of that rut, PV did very well.

In the mid-1980s, two 1-MW systems were built by the Sacramento Municipal Utility District (SMUD) and were jointly funded by DOE and SMUD. Work on subsequent SMUD projects was dropped by DOE when DOE budgets fell precipitously, and it was clear that emphasis was needed in R&D rather than for demonstrations.

ARCO Solar, a private company, also played a role in demonstrating PV on a large scale. They funded a project called the Lugo plant at Hesperia, California (1 MW). Then they built the largest demonstration to date, a PV field at Carrisa Plains, California. Rated at 6.5 MW_p, the system was installed in 1983–1984, taking only 1 year from blueprints to operation. Annual output is about 12 million kWh_{DC}. Today, it runs with only computer supervision and no on-site staff. During the 5 years it has been in operation, it has performed within 2% of predicted electrical output. Those predictions were based on the amount of sunlight available at the site.

At the time Carrisa Plains was built, there were substantial tax credits for solar energy in both California and federally, so Carrisa Plains could be regarded as another high-cost demonstration that depleted PV of funds needed for R&D progress on advanced technologies. Atlantic Richfield has never published the cost of Carrisa Plains, but assuming a $10/W_p$ system cost, it must have been around $70 million (before reduction for tax subsidies). Can one imagine how much more effective it would have been to have used the Carrisa subsidies to support emerging technologies like CIS or cadmium telluride, which together were receiving about $2 million annually at that time from DOE? Tax credits for projects that are not even marginally economical are a mistaken use of federal resources. The information gained, and the PR from the visibility of such demonstrations, do not even compare with the value of developing a PV technology that could eventually cost one tenth to one fiftieth as much.

Recently, the DOE joined with Pacific Gas and Electric, Electric Power Research Institute, California Energy Commission, and others to fund a *new* demonstration project called PVUSA. Begun in 1986, it was specifically designed to introduce new, innovative PV technologies via very small-scale projects. Such an approach does not reproduce the errors of the past, although the redirection of even the most modest DOE funds ($1–$2 million annually) needed for PVUSA has somewhat curtailed R&D progress at the cell and module levels for the same new technologies. To its credit, PVUSA has been eminently successful in introducing systems using new technologies such as a-Si, CIS, CdTe, and thin-film and ribbon silicon. Many of the systems have been the first ever use of these potentially lower-cost approaches. Data from them will provide a technical basis for improving the technologies tested or for gaining confidence in their use in PV products and systems.

PVUSA was designed to be a consortium of utilities with interest in PV. They partially fund new projects in order to gain firsthand experience with PV. PVUSA located its main test area in California, but new projects may be located around the US on land supplied by interested utilities. Interest in PV is growing among utilities, but as yet there is no wide participation in PVUSA. Their concerns are on more near-term issues such as dealing with their nuclear power plants and competing with new, independent suppliers of energy.

The question of using money for large-scale demonstrations versus using it for R&D is a complicated one. Some may correctly argue that the *government* would not have provided the money for R&D without using some or most of it to show off the results of that R&D. If the money had not been spent on PV demonstrations—with their limited but real value—it would have gone to non-PV uses. Still, even today (with PVUSA), we see that decision-makers will direct money away from R&D to demonstrations, even at a time of tight budgets.

It may be argued that PV is *now* ready for major subsidies because it is so much closer to competitive costs than it was late in the 1970s. The PV industry would be able to further reduce costs through major production increases (i.e., economies of scale). How tempting it is—at any given time—to declare technical success and spend a lot of money stimulating sales! The point is not that demonstrations are a mistake. The point is that *most true progress emerges from applied R&D*, and the need for R&D is often forgotten when people want immediate solutions.

Distributed Power

Although central station PV may always dominate distributed PV in the US, distributed PV will have its uses in the US. Globally, distributed PV could be more relevant to those countries that do not already have a utility grid. In the US, distributed PV would most likely be used while maintaining a link with utility electricity from the grid (see Figure 7 in Chapter 3). Backup batteries would be avoided by maintaining such a connection with the utility. PV electricity would be sold back to the utility during the day and then bought back at night or under cloudy conditions to supplement the PV system. Although the PV electricity generated by a residence is not typically of great use to that residence during the day (especially with the trend toward two-earner families), it is likely to be of great use to the utility, since it comes when commercial demands are at their peak. In addition, distributed PV electricity would be generated closer to where it is needed. Transmission losses would be much reduced for such electricity in comparison with electricity from a distant power plant. For instance, normal transmission losses average about 8%. But transmission losses as high as 20% can occur during periods of peak demand when power is sent to outlying consumers. Peak periods are frequently during summer days, when commercial users require electricity for air-conditioning. Thus, the transmission losses associated with these peaks could be much reduced by distributed PV, even if much of that PV were on the rootops of single-family homes.

To a utility, a PV array on a residential or commercial rooftop would look just like a conservation measure. Various loads would be diminished by the presence of PV. Utilities (and their controlling commissions) are coming to learn that there are few things as cost-effective as increased conservation. Greater regulatory support should be forthcoming for conservation, and for anything that accomplishes the same thing (e.g., PV).

A particular use of PV that closely parallels conservation is called peak shaving. Peak shaving entails the use of a PV array atop a commercial or residential building to provide electricity during daytime peaks of electricity demand. Such utilitywide peaks are usually during the summer for air-conditioning. Since PV power production is greatest at about the same time (because sunlight is also greatest), PV is very well matched to this valuable use. In fact, utilities charge much higher rates (up to 20 cents/kWh) for peak power because that power is the most costly for them to deliver using conventional units. The reason for this is that the units—typically gas turbines—are being used only a few hours a day, so their capital costs are relatively greater in proportion to the electricity that they produce.

PV arrays on the rooftop of a business with a high peak summertime demand can reduce that demand substantially, saving electricity expenses just when electricity costs the most. This can make PV cost-competitive far sooner than it would be if it were to compete directly with conventional electricity at off-peak times. The next generation of PV technologies, to be manufactured by plants currently in design or under construction, should be competitive for this substantial application (a US market of at least $1 billion/year). Much of the need for peak shaving is in the US Northeast, because that is where electricity costs are high and where brownouts are already occurring during the summer. The next generation of PV should be able to compete in the Northeast, where the high price of peak electricity (20 cents/kWh) will far outweigh the fact that there is about 25% less sunlight there than in the Sunbelt region.

Conclusions

Existing uses of PV are small but significant in that they have helped to perpetuate a PV industry, provided evidence that PV could be used dependably for a variety of purposes, and have helped educate people about PV's qualities and potential. But PV demonstrations have also drawn government funds and attention away from R&D efforts needed to support the newer but potentially more promising PV technologies. Sometimes demonstrations have been used as a substitute for research—i.e., as a sort of gesture to claim cost-effectiveness, when the opposite was true. If PV R&D had always been supported at an extremely high level ($150 million annually), such demonstrations would not have been a problem, especially if they had been done on a reasonable scale. But the opposite has been true: PV R&D has frequently been sparsely funded and large-scale demonstrations have had more than their fair share of funding.

The fault for this problem is shared by many. For instance, industry has often been behind the decision to build demonstration projects. Corporations see them as a fixed market for their products, something needed to keep them in business. In the past, the opposite may have been true—the nascent PV industry sold into a fixed market and then collapsed when the market disappeared. Corporations weren't stimulated to do the needed R&D to make their products cost-effective without subsidies. But another force behind the demonstration projects has been politicians, who may see them as more impressive, politically, than supporting ongoing R&D (which can seem endless, unrewarding, and much less showy). Somehow, there is an idea that making something big proves its applicability.

The various forces impelling demonstration rather than R&D are still present. In fact, they are growing along with the desire to show that PV can make a big impact on environmental problems like the greenhouse effect. It is quite possible, as we enter the 1990s, that our government will take the same simpleminded view as it did at the end of the 1970s and decide that PV has come far enough to implement it as is, despite the major cost reductions that a bit more patience and R&D would reward.

Government programs need to achieve a balance in which new technologies are amply supported while mature, but promising technologies are brought into production. Only those technologies that could eventually become cost-effective should be supported with market stimulus. At any cost, "last-year's technology" must not be bought with taxpayers' money. Taxpayers' money should only be used to stimulate progress. One way to do this, in terms of demonstrations, is to require that pricing levels for new systems drop, year by year, toward a fixed-target–true-cost competitiveness. Companies that fail to keep up would be eliminated. Achieving a balance in the 1990s between market support and research will be important for the long-term health of PV.

15 ☼ Future Options for Storage and Hybrid Systems

Much of the potential of PV depends on the successful development of systems that compensate for the intermittency of sunlight. These are likely to be in the form of electric storage or hybrid systems. Even today, PV is closely wedded to storage in the form of batteries or attachment to an electric grid. In the latter case, the grid acts as a sort of infinite storage. This is the form of storage, or compensation, that is expected to be crucial to the use of PV for peak power in the US around the turn of the century. Without the grid acting as backup, PV could not be used in the US for this need. Americans simply would not tolerate intermittent access to electricity. So even today, PV depends on storage, and it is a relationship that is likely to expand in years to come.

Throughout the utility industry, there is a growing interest in electric storage. This results from the ongoing trend away from installing new conventional generation capacity. This trend is based on several factors:

☐ Utilities engaged in near-catastrophic overbuilding of new generation facilities during the 1960s and 1970s. This was based on their predictions of ever-increasing demand for electricity. As usual, they took the view that the future would be a simple extrapolation of the past. Electricity demand was growing at 6%, therefore it would always grow at 6%. They missed the oil crisis and the conservation that followed. Demand growth was near zero for nearly 10 years. The result was a temporary glut of electric capacity during the 1980s.

☐ A risk-averse planning approach is now in place at utilities, partially because of the glut of capacity and partially because utilities were burned by their optimistic economic assumptions about nuclear energy. Fear of starting major new projects assures that capacity will only be added when a need is clearly evident.

☐ Conventional nuclear and coal plants take from 5 to 10 years to build. Other, quicker options become, in contrast, more attractive.

☐ Changes in regulations during the oil crisis allowed new competition in electricity generation from independent power producers. These exploit special opportunities, like cogeneration, to provide relatively inexpensive power in small increments.

☐ Environmental issues are restraining the siting of new coal or nuclear plants in most regions.

Because utilities are finding it difficult to build new conventional coal or nuclear plants, they are becoming more open to alternatives. No longer will they be adding a new baseload plant every few years. Besides alternative sources of energy (like PV), they are becoming more interested in electric storage. For them, electric storage would provide a means of storing cheap electricity from their conventional coal and nuclear plants at times when it is not otherwise needed, i.e., during off-peak periods such as late at night. This is the utilities' cheapest electricity—about 2–4 cents/kWh. They could store it at night and then sell it to the consumer at their peak periods during the day and early evening. Without storage, supplying these peak needs costs the utilities about 10–20 cents/kWh because they use expensive equipment for only a few hours a day to supply it. At other times, the equipment lies dormant, which means it has higher capital costs for every kilowatt-hour produced. By storing cheap, off-peak electricity, the utility can cut its costs significantly—if the storage method itself is economical.

Therefore, we may expect that utilities will expend much effort during the 1990s and the first decade of the 21st century developing inexpensive, efficient electric storage. Meanwhile, PV will also be maturing. Just when PV becomes inexpensive enough for storage to matter, economical storage should be available for PV to exploit on a large scale.

Various electric storage options exist, although none of them has reached a point where it might be regarded as truly economical. These, and a number of hybrid energy-production systems that act like storage, fall into the following categories:

- □ Batteries
- □ Pumped hydro
- □ Compressed air
- □ Flywheels
- □ Superconductivity
- □ Hydrogen production
- □ Utility grid

Batteries

Batteries are the classic choice for storage in existing PV systems. But for the long term, the present set of batteries is ill-equipped to be of real use for PV. The rechargeable batteries now in use for PV are lead–acid batteries and nickel–cadmium batteries. In most small PV systems, lead–acid *automobile* batteries are used despite many drawbacks. The problems with these batteries include that they are not particularly efficient (only 50% of the stored electricity can be reclaimed), must be replaced often (at least every 6 years), are relatively costly ($100/kWh of capacity for storing electricity), take up a lot of space, require much material (about 10 kg/kWh of storage capacity), and present some chemical safety hazards. In addition, they can withstand only about 400 cycles of being charged and discharged; and even a small array (1 kW) requires a surfeit of them: 70 automobile batteries to back up a system through 3–5 days without sun.

The cost for lead–acid batteries of about $100/kWh is the cost of producing 1 kWh when the battery is discharged once. The real cost of a battery depends on how many times it can be cycled. Suppose it can be cycled 400 times. In reality, cycling life depends on how much of the stored energy is being used during each cycle (called the depth of discharge) and on the ambient temperature. We are assuming 20% discharge per cycle. Then its cost per kilowatt-hour of electrical output is ($100/kWh)/400 cycles, or $0.25/kWh *based on only the cost of the battery*. Compared to the cost of conventional electricity (about 7 cents/kWh), this is way too high to be economical and is unlikely to be of use except where there is no other choice. On the other hand, if the same battery were capable of 2000 cycles instead of 400, it would add only 5 cents to each kilowatt-hour of production. Some lead–acid batteries designed for other uses besides cars (industrial trucks, marine uses) can cycle 2000–6000 times. Unfortunately, they cost much more per kWh of capacity than do car batteries. The upshot is that their energy cost is about the same.

How far do lead–acid batteries have to go before they become cheap enough to be of use for PV on a large scale? We can probably afford a storage system that adds less than 5 cents/kWh to electricity costs. This seems like a lot, but with DC PV costing only 3–5 cents/kWh, a system with storage at 5 cents/kWh would allow final costs of below 10 cents/kWh—which seems to be a reasonable cost for electricity in the 21st century.

To come up with a cost for stored energy, we must add the capital cost per kilowatt-hour to the cost of the feedstock electricity. However, that feedstock is not perfectly stored. Our final cost must take into account the losses during storage. We may expect that lead–acid batteries designed for PV will someday reach at least 70% efficiency. Costs are increased by 43% (1/0.7) by this 30% loss. Thus, if DC PV cost 4 cents/kWh, its cost would be 6.3 cents/kWh before adding the capital costs of storage. For those to be under 3.7 cents/kWh would require a 2000-cycle battery to cost about $74/kWh of capacity. One capable of 6000 cycles would be economical at about $225/kWh.

What about other kinds of batteries? Another familiar rechargeable battery is the nickel–cadmium battery. These currently cost about five times more than car batteries but can be cycled about 10,000 times. The net effect is that costs are about the same as for lead–acid batteries. However, their useful life (number of cycles) is not as affected by temperature or discharge, so in practice they generally cost less to use over the life of a PV system.

The amount of material within these batteries is large because they require large amounts of material to store relatively small amounts of electricity. This is also called their energy density, i.e., how much energy they can store per unit of material. Lead–acid batteries can store about 0.1 kWh of electricity in 1 kg of material. To store a large amount of electricity—say 20 days of all US electricity—would require over 10^{12} kg (a billion metric tons) of material. This is an impossibly large amount of material, especially if a portion of it is lead, nickel, and cadmium.

Other battery options are being developed. One of some interest is based on sodium and sulfur. These are batteries that operate at elevated temperature (about 300° C) but have the potential of reaching energy densities four times higher than lead–acid batteries. This would reduce the materials requirements associated with storing a large quantity of electricity. Also, sodium and sulfur are less problematic than lead, cadmium, and nickel in large quantities. The estimated future cost of these batteries is about half of lead–acid ($50/kWh), and they could be recycled almost 4000 times. As such, they would add only 1.3 cents/kWh to the cost of electricity. They are expected to be of about the same efficiency as lead–acid batteries, i.e., over 70%. If these kinds of performance capabilities could be reached, sodium–sulfur batteries could play an important role in electricity storage.

Pumped Hydro

The concept of pumped hydro is a simple one: water can be pumped uphill behind a dam and then used at some later time to produce electricity. Pumped hydro is already used by various utilities in parts of the country where dams are available for this form of storage. Efficiencies are about 70%. An advantage in relation to batteries is that dams have very long lives—50 or more years—so initial capital costs are compensated by great durability. Another, smaller advantage in relation to batteries is that hydroelectric power is produced as AC, so a pumped hydro storage system can double as a DC-to-AC inverter for PV electricity.

Where it is geographically favored (i.e., where hydro already exists), pumped hydro is already a cost-effective method of providing peak electricity. Off-peak power is used to pump the water uphill. However, the number of dam sites for this approach is limited. The only practical way to use pumped hydro for larger applications would be to develop a system that was not geographically limited. This would entail building a hydroelectric plant totally devoted to storage. Such a system is underground pumped hydro, i.e., a system in which water is pumped up from a deep (5000-foot), underground hard-rock cavern to a ground-level reservoir. Such a system would have very substantial capital costs, raising the overall electricity cost significantly above costs associated with using existing pumped hydro. However, even with these costs, various analyses have suggested that pumped hydro could meet our requirement of adding less than 5 cents/kWh to the cost of feedstock DC PV. Since the working fluid of pumped hydro is water, materials limits would not be an issue. Siting issues might be more likely to limit widespread use.

Compressed Air

There has been renewed interest at the Electric Power Research Institute and elsewhere in compressed air storage of electricity. The concept is simply to pump air into large, underground caverns and then use the compression later as a source of power to drive a generator. In existing systems, the force of the compressed air is supplemented by burning natural gas with it in a gas turbine. In the long term, the need for natural gas could limit the use of compressed air storage with PV (since natural gas is a fossil fuel) unless an alternative approach could be found. One might be storing and burning PV-generated hydrogen in place of natural gas.

An advantage of the compressed air approach is that there are many different caverns of the type needed for compressed air storage. Such caverns include deep rock caverns, salt caverns that can be emptied by pumping in water, or deep underground reservoirs of water under water-bearing sand or rock. About three-fourths of the US has these types of geological strata, so compressed air is a viable approach throughout the country. Compressed air storage is expected to be about 70% efficient. No materials limits appear to be of concern, so this method could be a storage technology used on a very large scale.

Flywheels

Flywheels are large wheels, usually made of metal or stone, that can store energy in the form of rotational energy. Energy can cause the flywheel to spin. Noncontacting magnetic bearings are used between the flywheel and its support to minimize friction losses. Flywheels are attractive for PV because a DC motor can power them and an AC motor can be used to remove the energy they store. As such they also act as an inverter. Flywheels can actually be quite efficient: 70–80%. However, they are limited in size by the great weights involved. Existing systems engineered for single-family homes can be 2 tons in size and spin at 15,000 rpm. Scaling to much larger sizes would be impractical and require massive amounts of material. In addition, costs have been quite high. Existing systems have about an 8-year life and add about 20 cents/kWh capital cost to the cost of feedstock electricity. The engineering problems associated with scaling up flywheels to very large masses or rates of spin suggest that this approach will have limited usefulness for storing major amounts of electricity.

Superconductivity

Superconductors are capable of storing electricity without any losses. Electricity can be fed into coiled superconductors and then withdrawn on demand. Existing superconductor technology, based on copper/niobium–titanium, is capable of achieving this feat but requires very low temperatures to do so. However, new superconductors based on materials such as copper–oxygen and yttrium are being developed to be of use at much higher temperatures. They are not expected to be available for actual systems until well into the next century.

The major elements of a superconducting storage system are massive superconducting coils suspended in a liquid helium bath, a cooling system for the bath, a support structure to hold the system, and a power-conditioner to control the movement of electricity in or out of the superconductor. The system must be built underground in order to resist major twisting forces induced by the flowing electricity within the coils. Another reason for underground placement is in case of a catastrophic failure and the release of all the energy in the superconductor.

A recent study performed by Bechtel National Incorporated, with support from Southern California Edison, analyzed the costs of such a system for various sizes of storage. The materials and construction costs for the coils and the support structure were 85% of the total cost for plants capable of storing 5000 MWh (1000 MW peak). For plants between 5000 and 10,000 MWh capacity, costs of \$150–\$130/kWh were estimated. The round-trip efficiency of the plant was about 94%. Of the 6% lost, 2.5% was from losses due to cooling the superconductor and the rest was from the inefficiencies of the power-conditioning equipment. A 20 year life was assumed for the system. Bechtel found that superconducting storage added about 5–6 cents/kWh to the cost of feedstock electricity, if that original electricity cost 6 cents/kWh. The hypothetical Bechtel storage system was used for about 5–10 hours a day to meet peak loads. This same storage system could be adapted to PV to smooth a very large PV system's output and to supplement it during the rest of the day and early evening.

The key advantage of superconducting storage is its high efficiency. Since inefficiency in storage leads to the direct loss of electricity, *all* storage costs scale with the inverse of the storage efficiency. This factor (1/efficiency) is the factor by which cost is increased (not including other costs, e.g., capital costs). The efficiency of superconductor storage is so high (94%) that this multiple is very small: 1.06. Other storage methods have lower efficiencies. For instance, a 50% storage efficiency means 50% of the electricity is lost, and that alone doubles cost.

For a superconductor storage system, most additional costs are capital costs that go up with the *size* of the needed storage, not the cost of the feedstock electricity. Because cost of feedstock electricity is not a major cost driver, one can visualize superconductor storage being an effective option for PV throughout PV's evolution from high to low cost.

Small, prototype superconductor systems are currently under development with funding through the US Strategic Defense Initiative. They use the existing, low-temperature superconductor technology. (In these systems, the stored electricity would be used for shooting down missiles in space.) Such systems are said to be over 90% efficient, which includes electricity used for cooling and other losses

needed for the reaction to proceed is reduced and used more efficiently. Usually, heat energy is less costly than electricity. Waste heat from another process could be a feedstock for the electrolysis. Or it might be possible to use cooling fluid from a hybrid PV/thermal system—or even combine a solar thermal and PV system—to drive an electrolyzer.

Coupling PV to an electrolyzer requires that the efficiency of the electrolyzer be independent of the output changes of the PV array throughout the day. Various theoretical and experimental studies have been done about this problem. They show that an electrolyzer can track changes in PV output with less than 7% losses. Thus, combined with an electrolyzer efficiency of 83%, the efficiency of using DC PV to produce hydrogen is about 78%.

What about the economics of an electrolyzer? At 10 MW capacity, the capital cost of a unipolar electrolyzer is about $170/kW. For comparative purposes, this is about 50% higher than we have assumed for power-related costs of a PV system—i.e., it adds less than a penny a kilowatt-hour to costs.

To come up with a total cost one must also take into account the 78% efficiency of the conversion process. DC PV costing 4 cents/kWh becomes 5 cents/kWh with this loss. Thus, hydrogen could be produced from 4 cents/kWh PV for roughly 6 cents/kWh. This is equivalent to about $17/GJ, where a GJ is a gigajoule—a billion joules. A gigajoule is very close to a million Btu (MBtu), a conventional thermal unit. A ten-gallon gas tank stores about a GJ of energy, so at about $17 per tank, we are in the right ballpark. (Other factors affect the relationship with gasoline costs—efficiency of hydrogen burning is much higher, for instance, which helps the economics. See below.)

One other method of producing hydrogen from DC PV is also of interest: using fuel cells in reverse.

Fuel Cells

Fuel cells are like batteries in that they are electrochemical devices. During normal operation, fuel cells depend on the continuous addition of a fuel—hydrogen or hydrocarbons—to produce electricity. Fuel cells were originally developed for use in space, where they were designed to use hydrogen and oxygen.

The hydrogen and oxygen are not burned within the fuel cell (as they would be in a turbine). Instead, chemical reactions of the oxygen and hydrogen at separate electrodes induce a flow of current that transfers charges through an external circuit (Figure 46). The hydrogen electrode can be made from copper, silver, nickel, nickel boride, or carbon. A porous carbon material with platinum, silver, cobalt, copper, or nickel oxide is usually used for the oxygen electrode. DC electricity is produced as hydrogen and oxygen are pumped through the cells. Water vapor is the product of the reaction, so hydrogen/oxygen fuel cells are environmentally compatible with PV.

Fuel cells can be run in reverse. That is, electricity fed into a fuel cell can induce a reverse reaction in water, splitting it into hydrogen and oxygen. Thermal energy can be added to the system to supplement the electricity, raising the yield of hydrogen per kilowatt-hour of electricity. Efficiencies over 90% are possible, although such units are not yet generally available.

Fuel cell costs are dropping rapidly as their development and

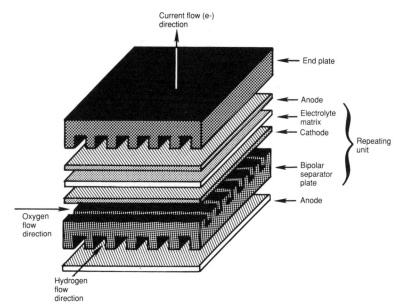

Figure 46. The basic structure of a fuel cell. Many units are stacked together to increase power output. Hydrogen and oxygen are fed into different segments of the fuel cell and react to form water vapor (H_2O), with the release of two electrons.

commercialization continue. Costs of about $0.5–$1/W are expected in the 1990s. At present, fuel cells are rather short-lived and have higher operations and maintenance costs in comparison with electrolyzers. However, rapid progress in improved dependability is also expected.

The costs of using fuel cells in reverse to produce hydrogen are likely to be about the same as those for using electrolysis. Reverse fuel cells are not yet commercially available. If they can be developed, they would provide a valuable capability: making hydrogen when the sun shines, making DC electricity from hydrogen when it does not. This dual role could reduce overall system costs.

Gas Storage

Various scenarios for the use of hydrogen fuel suggest that hydrogen and oxygen storage will be needed. Even if they are to be piped elsewhere for immediate use, smoothing the flow of gas is needed for economical transmission. For that, backup gas storage is necessary. If on-site use for continuous electricity is the application, significant storage is a necessity.

The hydrogen and oxygen can be compressed and stored. Suitable methods are similar to those associated with compressed air storage (see earlier in this chapter), with rock caverns, aquifers, or salt caverns providing appropriate storage cavities. In this case, the stored energy is much greater per unit volume than merely storing air, since the hydrogen and oxygen carry thermal as well an pneumatic energy potential. Both can be retrieved.

Ogden and Williams considered the cost of compressing and storing hydrogen in depleted gas wells in the Southwest. Such cavities can be charged with about 600–750 pounds per square inch of hydrogen gas. The gas wells are typically quite large and could be used as backup for a billion-watt PV system. They would add only about 0.1 cent/kWh to the cost of hydrogen/electricity. However, if a natural gas well were not available—i.e., for more general use—costs for storing hydrogen and oxygen would mimic the costs associated with compressed air storage. Much higher capital costs would be expected and thus higher storage costs.

Two other hydrogen storage methods are also possible: liquefaction and storage in metal hydrides. Although liquefaction could add several cents per kilowatt-hour to the cost of electricity, it would remove geographical limits. Large, on-site storage tanks would be enough to store billions of watt-hours of electricity (the energy densities of liquid hydrogen and gasoline are similar). Metal hydrides are also an option. These metallic materials have the ability to hold hydrogen molecules in a weakly bound state. They release the hydrogen as they are heated. For vehicles, they are problematic because they are heavy, but for a stationary use, they may be very attractive.

Continuous DC from PV/Hydrogen

For some applications such as major industrial uses, DC electricity is actually preferable to AC electricity. PV can be used in its natural state to supply DC electricity. The natural backup would be hydrogen fuel cells (which also supply DC). The PV hydrogen for the fuel cells could be made by electrolysis or in a reversed fuel cell.

Used in their normal, forward configuration, fuel cells produce DC electricity. Fuel cells can be about 70–90% efficient in converting hydrogen and oxygen to electricity. If one assumed PV/hydrogen costing 6 cents/kWh, then the inefficiency of the fuel cell would raise costs to about 8 cents/kWh, assuming that the fuel cell conversion efficiency would be about 70%. This is a conservative estimate: higher fuel cell efficiencies and use of the pneumatic potential of the storage are ignored. To come up with a total cost, the capital cost of the fuel cell must be added as well. In the system in which the fuel cell is being used as the electrolyzer, the fuel cell can be used to store energy when the PV array is producing electricity; and it can be used to produce energy when the PV array is off. No overlap exists between the two states, so no extra capital cost is needed. The capital cost of fuel cells in the late 1990s should be below $1/W. This would add another 4–6 cents/kWh to the cost of electricity; total cost would be about 12–14 cents/kWh. Further, the O&M costs and replacement costs of such a system would be greater since the fuel cells would be used almost all the time.

Many consumers would find continuous DC attractive. The characteristic owner of such a system would be generating DC PV on-site, using much of that power immediately for DC motors and other DC uses (almost all major applications can use DC) and storing hydrogen for when the sun is not out. As we will see below, some or even most of that hydrogen could be used as fuel rather than to make electricity—or both, via cogeneration schemes in which excess heat is also used. The local use of DC provided by PV would have two other advantages: it would eliminate the losses associated with transmitting AC electricity from a distant source; and it would remove the need for an AC-to-DC rectifier at the plant site. A rectifier is a costly item used to change AC to DC for DC uses.

Continuous AC Electricity

Stored hydrogen can also be used to make AC electricity. Instead of a fuel cell, combustion turbines can burn the hydrogen with the stored oxygen, in much the same way natural gas is burned in gas turbines. The procedure automatically results in AC electricity, eliminating the need for an inverter (if AC is the desired kind of electricity). The combustion product of burning hydrogen and oxygen is water vapor. No NO_x is produced in such a system because air is not present, i.e., no nitrogen is available to react and produce an undesirable emission. As such, the use of turbines with hydrogen and oxygen is environmentally benign.

The key to this option is the efficiency of the turbine. Efficiencies of typical steam turbines are quite low—about 30%. Natural gas is burned. A heat exchanger allows the transfer of thermal energy to water, which produces steam. Steam drives a generator. Although the natural gas burns at 1500° C, the steam is produced at only 550° C. This is the source of the low efficiency. In contrast, hydrogen and oxygen burn at about 2700° C. Although turbines cannot withstand this temperature, D. A. Mathis of Noyes Data Corporation suggested in a recent study that water-cooled turbine blades could withstand about 2200° C. Water injected in the superheated steam could be used to cool it to this level. The resulting efficiency from such a high temperature would be over 60%. In their report, Ogden and Williams chose 65% efficiency for this approach.

Besides the higher temperatures available to these H_2–O_2 burners, another reason that they are more efficient than conventional turbines is that they do not carry the weight and combustion losses associated with burning fuel in air. Air has weight and carries away heat. Finally, the hot gases emerging from the H_2–O_2 burner can be condensed into water and the extra energy thus liberated could also be extracted.

The turnaround efficiency between feedstock DC PV and continuous AC using hydrogen storage and a turbine would be low— about 50%. However, the added cost/inefficiency of an inverter would be eliminated, since turbines produce AC. Suppose all the PV-generated electricity were used to make hydrogen. Then the cost of continuous AC electricity, based on DC PV at 4 cents/kWh, would be about 12 cents/kWh. Gas turbines cost about \$400/kW, which is a substantial cost penalty (about 3 cents/kWh). However, this is an overestimate of cost, since in most PV applications, the AC would not be all coming from hydrogen. The PV system would be supplying much of its electricity directly (through an inverter) and storing some fractional amount. In that case, some proportion of the system would be a normal, intermittent AC system costing about 4.5 cents/kWh. Averaged with the higher cost electricity coming from storage, the total system cost would be closer to 9 cents/kWh (for about 50% storage).

Hydrogen Economy

Perhaps the ultimate use of PV would be to produce hydrogen for the hydrogen economy. Intermittent DC PV feedstock would be used to split water. The gases would be compressed, stored, and then piped or otherwise moved to where they would be used as portable fuel. We have already seen that the efficiency of converting DC PV to hydrogen is quite high: 78%. PV at 4 cents/kWh would produce hydrogen via electrolysis at 6 cents/kWh (\$17/GJ).

Existing natural gas pipelines could be used to transmit hydrogen nationwide if they could be modified to avoid a phenomenon called hydrogen embrittlement. Embrittlement is caused when the small hydrogen molecules diffuse along grain boundaries in steel pipes, causing such pipes to leak. However, several coatings have been found that could ultimately be sprayed through pipelines to protect them from embrittlement.

If these preventive approaches are not practical, improved pipelines devoted to hydrogen could be built along the rights-of-way of existing pipelines. Ogden and Williams estimated the cost of transmitting gas a thousand miles to be about $1.90/GJ (0.7 cent/kWh). They assumed that new pipe was built at $0.35/GJ and hydrogen was compressed to 1800 psi, costing about $1.5/GJ. Thus, transmission would raise the cost of PV/H$_2$ by 10% to about $18.70/GJ.

The uses of hydrogen are potentially ubiquitous. The major one would be for transportation to replace gasoline or some future alternative such as methanol. The latter can be processed from hydrocarbons or from vegetable matter called biomass. Most energy planners expect that methanol will someday displace gasoline as the preferred transportation fuel. It is mentioned extensively as a substitute in new clean air legislation. However, methanol produced from coal causes massive carbon dioxide emissions and requires vast water resources—which are not always available where they are needed. Methanol from natural gas could be a transitional fuel; but using it as a replacement would strain natural gas reserves beyond their expected limits. Thus, most experts believe that the ultimate source of methanol is likely to be biomass. Biomass can be processed into various alcohols, including methanol.

However, PV/hydrogen has several critical advantages over biomass: it requires much less land and much less water. Biomass is about 0.1–1% efficient; PV can be more than 10–20 times more efficient. Calculations of the land area needed to provide all of our transportation needs from biomass-derived methanol suggest that most of our nation's cropland or forests would be needed. As a corollary, the land needed for biomass must have access to plentiful water supplies. Biomass uses about 25,000–60,000 liters of water per gigajoule of energy; PV with hydrogen needs only 63 liters, about 10,000 times less. PV/H$_2$ is not water limited, while biomass is very much so.

Hydrogen has some drawbacks, however. One is storing enough of it in a car to provide a reasonable range between fill-ups. Fuel efficiency will be critical to the hydrogen car option. Preliminary analysis shows that hydrogen can be about twice as efficient a fuel (per GJ) as gasoline. This means it can be twice as expensive per gigajoule and still be economical. Another issue is a perceived problem with safety. Old newsreels of the crash of the German hydrogen blimp, the Hindenburg, cause people to fear hydrogen. Actually, most of the passengers on the Hindenburg survived, which would not have happened if they had been in a conflagration of natural gas. Among the technical community it is well known that hydrogen is a safer fuel than either gasoline or natural gas.

When a practical means for providing fuel is needed in the next century, the combination of PV and hydrogen will be one of the choices with the best economic and environmental bases.

The Electric Utility Grid

As it does today, the electric utility grid may continue to be a critical part of hybrid PV systems. After all, the conventional power on the grid is the basis for our early uses of PV. But in the future, the grid's role could be radically different: it could act as a supplement to PV rather than vice versa. The utility could act as the central means of storing PV (and other) electricity and then distributing stored electricity when PV is not available. This would be especially true for winter electric demand, when PV production would be about 30 to 80% of summer production. The major part of the stored energy could be from PV, so the utility would have a minimum of actual generation capacity.

Whether energy would be stored as hydrogen or within superconductors or in some other form is unclear. Some combination seems most likely. For instance, superconductor storage, with its high capital cost per stored kilowatt-hour, could be used for smoothing daily loads. On the other hand, hydrogen production could be used to meet seasonal variations, since capacity costs (cost of gas storage, i.e., $/kWh, and other capital costs) are a much smaller proportion of cost. The utility grid (with storage) would act as a national—perhaps international—leveler of supply and demand, providing electricity from where it is being made to where it is needed.

Conclusions

The purpose here is not to rationalize any specific storage or hybrid system, be it superconductivity or the hydrogen economy. The same is true for transportation: eventually some nonpolluting means of providing energy in a portable form will be developed. PV hydrogen is a good possibility, but we have not even mentioned battery storage for cars. Obviously that would be another major possibility for using PV-

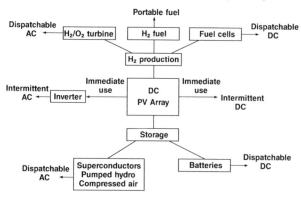

Figure 47. There are numerous storage possibilities for PV electricity. Their development during the next 10–30 years will have a major influence on the potential of PV to be used to meet a significant portion of our energy needs.

provided energy to avoid known environmental stresses. Similarly, we cannot yet visualize how an interactive grid–storage system would really work, but we know could contribute to its effectiveness.

The real purpose here is to describe some of the numerous storage and hybrid systems that could be useful to the future of PV (Figure 47). With storage, PV could make a large impact on all of our energy needs. PV is a nearly unique source of electricity. When it is low enough in cost, we will surely become very clever in how to use it. The suggestions of this chapter will likely pale before the elegance and efficiency of real systems.

16 ☼ The Future
Technology Issues

Even though most people do not even know PV exists, its future is already assured. PV systems work, producing electricity reliably over extended periods (up to 30 years or more). Although cost is still too high (about 30–40 cents/kWh), the plants now in the design phase for several technologies (ribbon silicon, a-Si, cadmium telluride, copper indium diselenide, concentrators based on silicon or gallium arsenide) should make PV modules that will produce electricity at 10–20 cents/kWh (or lower) in the mid-1990s. The basis of these cost reductions will be improvements in technology and significant economies from mass production.

Meanwhile, progress in PV R&D continues. Every year, significant problems are overcome, cell efficiencies are increased, and module sizes and efficiencies go up. All this simply recapitulates the facts made plain within the bulk of this book: PV exists, and it will someday be inexpensive enough to make a significant contribution to global energy.

Behind the progress of PV as a technology is a loyalty to PV as an idea. Technical people who are familiar with the true characteristics of PV—the low costs of its component materials, the potential for very inexpensive manufacturing—have strong belief in its potential success. This has given PV a special strength. Through enormous fluctuations in public interest, in government funding, in corporate commitment, PV technology has continued to progress. Even when one or another company folds, and a particular PV technology seems threatened, new companies emerge and take the best ideas of the old ones a step further. Some big oil companies have dropped out, and small entrepreneurial companies have sprung up to take their places. Often, the process has had a silver lining. The new companies shed the rigid, often conservative ideas of the previous companies and take off in directions of greater eventual payoff. Technical people start new companies because they have better ideas than the management of their previous companies. PV has had the quality of a phoenix, continually renewing itself and never succumbing to the many blows and setbacks along the way. Behind the enthusiasm of the people of the PV community is the idea that PV will work and be of great value.

Nonetheless, this does not mean it will work immediately or that the United States will have a major PV future. Too many other things influence those outcomes.

One aspect of the problem is technical. Issues remain for all the PV technologies, at least in terms of reaching their full potential (i.e., cost of about 4 cents/kWh). Now we will discuss those technical issues.

Crystalline Silicon

Throughout the history of PV, the steadiest and most reliable PV technology has been crystalline silicon. Modules made from crystalline silicon have been the largest fraction of those sold, especially for the remote power market, which is the most important and fastest growing (30–40% annually) PV market. No other technology can say that it has in any significant way displaced crystalline silicon as the quintessential power module.

Crystalline silicon is also the most reliable PV technology, with a 30-year outdoor life now proven. The issues that remain for crystalline silicon are efficiency and cost. Many researchers have been able to demonstrate exotic efficiencies with crystalline silicon (over 20%), but in most cases the structures necessary to achieve high efficiency have not been made compatible with low-cost manufacturing processes. Module efficiencies are in the 11–13% range for modules available to the public; research modules over 15% remain a rarity. Much work needs to be done to transfer the cell design advances of the last decade (front and back-surface passivation, texturing, ultrahigh-quality crystal growth) to practical silicon production.

On the cost side, new processes based on ribbon growth are reaching maturity. Manufacturing plants are being designed by several companies, including Mobil Solar and Westinghouse. These facilities should lower the cost of silicon modules considerably while maintaining efficiencies in the 11–14% range.

The feedstock material for silicon cells, semiconductor-grade silicon, is now available at $20–$30/kg. Continued improvements in the economics of purifying raw silica should keep this cost low and could eventually reduce it to about $10/kg. Meanwhile, various groups (Westinghouse, Astropower) are finding ways of using less silicon in their modules, so that materials cost on an areal basis could approach the costs associated with thin films (i.e., under $10/m^2). It may even be possible to mimic the thin-film approach by fabricating relatively thin (20 micron) silicon on large-area substrates and then finding ways to monolithically interconnect the cells.

As the various manufacturers of silicon scale up the size of their facilities, they will naturally achieve economies of scale. Processes that are now done by hand or with minimal automation (cell interconnection into modules, framing, wafer handling) will be fully automated. Capital and operations and maintenance costs for robots will replace labor costs. Very substantial cost reductions will naturally occur.

The silicon technology has many strengths: low-cost feedstock, a strong base in fundamentals and processing (including knowledge drawn from the computer industry), proven reliability, and proven efficiencies over 20%. Research to address the remaining cost and efficiency issues can build on these strengths. Silicon is a sound technology that will continually drop in cost. It remains the lowest-risk PV technology.

Amorphous Silicon

Many people familiar with the development of new technologies have described an evolution (called a learning curve) in which there is a very pronounced dip in enthusiasm that occurs just prior to real success. For instance, new technologies (room-temperature supercon-ductors, fusion, PV in the 1950s) are greeted with unbridled excite-ment. The potential is clear, and some vague idea exists of how to make the potential real. But the usual progression is that much work follows, some success is achieved, and then many of the substantial difficulties in the way of full success become obvious. In order for those technologies to succeed, someone must stick to it through this dark period. Eventually the technology makes a real contribution, although it may not be the one first envisioned.

a-Si is now in its dark period. Enough effort has been devoted to it that real a-Si products exist, and real systems are in place. But the remainder of the initial enthusiasm about a-Si cannot disguise the fact that warts are showing. a-Si has a long way to go in terms of the three critical parameters defining the success of PV: efficiency, cost, and reliability outdoors.

The instability of a-Si to exposure to sunlight has driven the technology in several key ways. First, in the early 1980s it was unclear as to whether (1) some physical aspect of a-Si could be adjusted to rid the technology of its sensitivity to sunlight; and (2) if not, it was uncertain whether the instability would be fatal: Would the observed initial degradation of about 20% in the first month outdoors simply continue until the devices were useless?

Neither of these questions has been fully answered, although much progress has been made in both areas. Most progress has been engineered, in terms of ridding the technology of instability. It was found early on that thinner a-Si layers resulted in a substantially slower rate of degradation. The result was that groups found ways to make thinner cells in both single- and multijunction designs. Light-trapping schemes (back reflectors) were developed to allow layers to be thin without loss of light absorption and current density. New, a-Si alloys (with germanium and tin) were designed to absorb lower-energy light to optimize multijunctions.

Progress in these new designs, as well as parallel work in module design and encapsulation, is starting to answer the second concern: will the instability be a fatal flaw in actual devices? The answer seems to be no. As the newer devices based on thinner layers are included in modules, the instability appears to be reduced. Rates of degradation are becoming slow enough that competing effects can dominate them.

For instance, it was found that heating a-Si devices can reverse degradation. Even at outdoor operating temperatures (about 60° C), the degradation within the a-Si can be reversed. In actual use outdoors, this anneal can counteract the light-induced degradation. Added to this, the rate of degradation has been substantially reduced, so that the thermal healing of the cell can actually fully counterbalance it. Complete data are not yet available. Proof does not yet exist that the light-induced degradation is under control. But logical arguments exist supporting this thesis, and expectations are that the losses can be kept to about 20% over the life of a module.

Another way in which instability has driven the a-Si technology is in terms of the issue of achieving a reasonable efficiency. a-Si modules must reach 18% efficiency if they are to be stable at 15% efficiency, the long-term goal of the technology. Today's stated efficiencies—8–10% for experimental modules and 5–7% for those available as products—are not stable, postdegradation efficiencies. Those are about 20% lower. The upshot is that stable module efficiencies are low: about 4–5% in the field, only 8% in the lab.

To reach any kind of acceptable efficiency, the a-Si technology has been driven to adopt a more complex approach—multijunctions. This means that new, lower-band-gap a-Si alloys had to be developed; or other materials, e.g., CIS, adopted for use with a-Si. Along with the more complex structures have come increased processing steps and potentially reduced yields. These are burdens on the future of a-Si.

But good progress in efficiencies is being made in the lab. Multijunction a-Si-alloy cells made by Energy Conversion Devices have a measured efficiency of over 13%. Hybrid a-Si on $CuInSe_2$ cells of 14.7% has been fabricated as well (ARCO Solar). The next step will be to transfer these lab-scale results to module production. The achievement of 15% modules remains a valid goal for such multijunction a-Si structures.

Cost is also an ongoing concern. a-Si by glow discharge is a relatively slow, vacuum process. Actually, the slowness of the a-Si deposition process is becoming less of a problem. Rates are increasing; meanwhile, layers are getting thinner, and the slow rates allow for the good control needed to make very thin layers. Still, it takes time to make several thin layers in various vacuum chambers. Substantial capital investment is needed to build a manufacturing line from high-priced vacuum equipment. The resulting line is made less flexible to adjustment and improvement because of the substantial investment required to change it. But manufacturers claim that at the annual production rates that they project for increased PV markets, economies of scale will be significant. Production costs will resemble materials costs and should be low. Actually, their argument is very similar to the argument promulgated by the crystalline silicon industry, which also has capital-intensive processing. But advantages of a-Si in relation to crystalline silicon include the lower materials costs of the thin-film approach; and the fact that cell interconnection is done as part of the module fabrication rather than as a subsequent labor-intensive (or robotics-intensive) step.

A final issue is safety during manufacturing. Several toxic gases (e.g., silane, germane, diborane, phosphine) are required for making a-Si materials and alloys. But manufacturers are confident that procedures will be in place for controlling these gases and minimizing risk. These gases are already used in large quantities within the semiconductor industry. Once completed, a-Si modules are considered fully inert and without any further environmental or safety risks.

Research continues in the area of stability. Some hope remains that a-Si can be stabilized completely. Meanwhile, engineering approaches have progressed enough to suggest that the problem can be controlled even without such fundamental progress. New multijunction designs are progressing, raising efficiencies in the lab and holding promise for the future when they will be incorporated in modules. Also, the knowledge gained by developing multijunctions is adding to the sophistication and depth of the a-Si technology. Higher processing rates are being achieved, and new processes are being looked at. Successful consumer products have been introduced, but power modules remain a goal for the future. A number of companies are sticking with a-Si through this flat period on the learning curve. They feel strongly that they will emerge with a successful, competitive technology.

An advantage of both a-Si and crystalline silicon is that they depend on silicon: a plentiful, nontoxic, well-understood material. The remaining PV technologies (copper indium diselenide, cadmium telluride, and gallium arsenide) do not have these advantages. This issue alone—use of nonsilicon materials—has kept interest in these technologies to a minimum outside of the US.

Copper Indium Diselenide

Copper indium diselenide ($CuInSe_2$ or CIS) has been a technology that has received a lot of attention recently. In the last two years, CIS has achieved efficiencies that put it at the top in relation to all other thin films: 14.1% cell efficiency, 11.1% 1-ft^2 module efficiency, and 9.7% 4-ft^2 module efficiency (1990). These are all records. In fact, they are substantially higher than any other competing thin film. Meanwhile, CIS modules tested outdoors at SERI for almost a year and a half have been stable within a 3% measurement error, which is the best result for any thin film so far. Because of these factors, CIS has been propelled into the lead among the thin films, at least as far as its perceived potential to reach ambitious module efficiencies (e.g., 15%) is concerned.

Yet issues remain. In terms of efficiency, there is a need for continuing progress. Cell efficiencies of about 18% (active area basis) will be needed for modules to reach 15%, because there is a loss of active area in modules as compared to cells. This will stretch the single-junction CIS technology nearly to its limits. Reduction of voltage losses (recombination in the junction and bulk CIS) will be needed, as will continued optimization for very high currents. Some effort to buildup the fundamental understanding of CIS will be needed, since this has been a weak point of the technology.

Stability also remains a concern. First, it has not yet been established if CIS is really stable. We know that it is sensitive to a relatively low-temperature (200° C) anneal—which in this case is used to *improve* its efficiency. However, subsequent overannealing begins to degrade performance. Is this equivalent to leaving CIS modules outdoors for very long periods at operating temperatures of 60° C? Will modules begin to degrade after several years in the field? Second, modules of any new technology have engineering bugs. We do not know if CIS modules are stable in terms of long-term effects of humidity. Are the metal interconnects durable? Is the CIS/molybdenum adhesion reliable? Even with the best results, it will take some time before CIS modules will be regarded in the same way as crystalline silicon modules in terms of reliability outdoors.

Cost is perhaps the major issue. Present processes for making the highest-quality CIS are rather costly—perhaps of the same order as a-Si processing costs. Two costly steps are based on sputtering metals:

molybdenum, then copper and indium. Although sputtering is a proven, large-scale, rapid-rate process, it is capital-intensive and does require a vacuum. Another process that may be costly is one of the final ones: application of a transparent conductive oxide, ZnO. Two existing choices are sputtering and a chemical vapor deposition (CVD). The sputtering approach is a different type from the sputtering of metals, and far slower. The CVD requires expensive feedstock materials. Both are vacuum processes.

But early work is under way to replace all of these processes with inexpensive, nonvacuum processes. Methods of spraying are being investigated for each of the materials, although no state-of-the-art efficiencies have yet emerged. Confidence is high that such procedures will be developed. If they can, they have the potential to halve the cost of making CIS modules.

A related matter is the cost of materials, especially the indium for the CIS layer. We know that indium is present in the Earth's crust in the same concentration as silver, which is mined at 8000 MT (metric tons) annually (many times more than we would ever need, no matter how large indium's use for PV). But we also know that indium supplies are small and indium is a by-product, dependent on the processing of primary metals. The problem can be ameliorated with thinner CIS layers, gallium substitution for indium, and improved indium utilization, all of which are being studied and should progress satisfactorily. The crunch may come if and when CIS reaches large production volumes—perhaps 1–5 GW annually. Then we shall see whether indium can be processed from other ores, and at what price. The indium in CIS costs about \$2/m² at present costs. A fivefold increase in indium costs could probably be absorbed without harming overall economics, especially if thinner layers with more gallium were produced. Still, the issue of indium availability is not likely to disappear.

Safety and environmental stresses are other issues. Most of the feedstocks for making CIS devices are relatively acceptable, with the exception of hydrogen selenide. The latter has similar problems (accidental catastrophic release of gas) to the toxic gases used in a-Si manufacturing. Much work is being done to avoid the use of hydrogen selenide (using selenium instead), but most manufacturers feel that replacement is not essential. If used in large production, hydrogen selenide would be produced from compounds containing hydrogen and selenium on-site and on demand; very little would be stored as inventory. This would minimize risks substantially and eliminate the need to transport the gas via public roads.

The environmental stresses associated with CIS modules are uncertain. CIS has not been classified as a toxic or hazardous material. It is such a novel material that it is simply unclassified. But its unclassified status will change, and we do not know what the result might be. In general, CIS will probably be subject to many of the constraints on disposal that cadmium telluride will have. As such, care will be taken that it does not enter the environment, and its disposal will be controlled. Materials will be recycled, perhaps paying for most of the disposal costs.

The CIS technology is in an ascendant phase. Its progress has been substantial despite a funding level that has been about 5% that of a-Si (which, in turn, has been 5% of what has been spent on crystalline silicon). For rapid progress on minimal resources, CIS is hard to beat.

But some issues remain. The technical ones (efficiency, cost, stability) are, if anything, rather minor. Certainly, they are acceptable within the traditional experience of such things in PV. Very similar issues in CIS and other technologies have easily been surmounted. Disposal of CIS modules could become an issue. The indium availability issue is perhaps more threatening for the long haul. If CIS is uniquely successful and can produce very low-cost electricity (3–5 cents/kWh) as it promises, will we have enough indium at a reasonable price to allow for a global impact? The numerous avenues toward ameliorating this issue (thinner layers, replacing some indium with gallium, new indium sources) suggest that pessimism on this subject is premature. Right now, CIS appears to be one of the best hopes for the future of PV.

Cadmium Telluride

Another thin film that has not received much overall support (about the same as CIS) but which has made significant recent progress is cadmium telluride. Efficiencies are good but not as high as CIS: modules are at 7.3% (1 ft²), cells at 12%.

Perhaps the unique aspect of cadmium telluride — and the reason greater interest is accruing to it — is its potential for very low cost.

Almost all aspects of cadmium telluride manufacture can be done without expensive, slow, vacuum processes. Examples are the spraying or electrodeposition of cadmium telluride. Others are the dip-coating of CdS, and the spraying of tin oxide. Even near-term facilities for making cadmium telluride in rather small quantities (under 10 MW) should provide low-cost modules (well under $100/m^2), and they should have great flexibility in terms of production modifications as the technology improves. The future also looks good: moderate production capacities (10–30 MW) should be enough for economies of scale to allow for very low-cost modules (under $50/m^2) priced at well under $1/W$_p$.

This kind of cost should allow cadmium telluride manufacturers to tolerate rather low near-term efficiencies, i.e., under 10%, and still achieve very low electricity costs (near 10 cents/kWh). For the longer term, the efficiency potential of cadmium telluride is expected to be about the same as that for CIS. In fact, cadmium telluride has a more optimal match to the solar spectrum, so its chances of reaching 18% cells and 15% modules should be better than those for CIS. Thus, although cadmium telluride presently trails CIS, scientists are optimistic that the situation is temporary. Like CIS, it should be able to make 15% modules with single-junction designs.

However, other obstacles must be overcome. One crucial hurdle is the question of instability. Cadmium telluride devices have a mixed reputation in terms of stability. Cells left unused for many years (a decade in some cases) remain unchanged. But cells used without encapsulation can quickly degrade, perhaps because of diffusion effects associated with the cadmium telluride contact. New designs avoiding cadmium telluride/metal contacts have been developed. And some encapsulation schemes seem to ameliorate or even remove the issue.

Three manufacturers (Matsushita, Photon Energy, and British Petroleum) claim that their cadmium telluride modules are stable outdoors. But SERI can only confirm the claims of Photon Energy. For the others, SERI simply does not have access to the data and the test conditions. Photon Energy's cadmium telluride modules, which we have tested, have not degraded for two-thirds of a year. The issue of cadmium telluride stability remains a concern, but reasons for optimism exist.

Production safety and subsequent environmental stresses are of particular concern with CdTe because of the presence of cadmium. Plant safety is less of an issue. Existing manufacturing approaches use 90–99% of the Cd; the rest is solid waste and easily handled. In addition, cadmium is a chronic rather than a toxic problem as far as plant safety; as such its releases are much more easily monitored and corrected than those of toxic gases. Manufacturers actually consider the handling of Cd and CdTe production as less challenging in terms of safety than the production of other thin films.

Product use and disposal are relatively more difficult issues. In a worst-case scenario, regulations barring unregulated, small-scale uses of cadmium telluride panels may become a serious barrier to their use on residences. Although problems with the small amounts of cadmium involved would be minimal and actual risks (e.g., with fires) unlikely, regulatory actions may be disproportionately large. Cadmium is a much-feared element, and PV is unfamiliar.

For larger uses, where utility control will help manage risks, the issues center around decommissioning and disposal. As stated in Chapter 11, total recycling of the valuable, highly pure cadmium and tellurium from the modules should allow this to be done at reasonable costs. In this way, the cadmium used in cadmium telluride will not enter the waste stream. In fact, cadmium otherwise released inadvertently from the production of zinc, burning of coal, or phosphate fertilizer could become the feedstock of cadmium telluride modules, thus reducing the overall entry of cadmium into the environment. Cadmium telluride modules could be part of the solution to the cadmium hazard, not the problem.

Some issues have been raised concerning tellurium. These have been expressed in terms of availability and possible environmental hazard. However, tellurium production is about three times that of indium, so concerns about availability are less than for indium. Similar strategies—thinner layers, substitutions—could also ameliorate the problem. As far as environmental concern, Te is much less dangerous than Cd. The strategies being adopted to minimize Cd hazards in the workplace would simultaneously minimize those from Te.

The CdTe technology is presently making a large impact on PV, especially because of its potential for very low-cost manufacturing. New groups are entering the field. However, two issues remain: the stability of CdTe modules and the use and fate of Cd as an environmental hazard.

Concentrators

Despite a number of idiosyncratic problems (nonuse of diffuse sunlight; geographic limits to cloudless regions), concentrators remain a viable and exciting option. They have the unique ability to offset costly cells with relatively inexpensive hardware (e.g., lenses, module housings). Exotic efficiencies well over 30% have been achieved with concentrator cells, bringing added interest to this field. In the meantime, an ongoing competition is raging between silicon and gallium arsenide-based cells to decide which will dominate in the long term. Issues related to concentrators include:

□ Progress in silicon and gallium arsenide efficiencies
□ Progress in low-cost cell production
□ The need for low-cost, efficient lenses
□ Costs associated with other module parts
□ Stability
□ Cost and reliability of precise, two-axis trackers

Progress in both silicon and gallium arsenide single-crystal cells for concentrators has been exceptional. Gallium arsenide cells are way ahead of silicon (over 30%—perhaps as high as 34%—versus about 28% for silicon), but gallium arsenide cells are more expensive as well. However, the cost of gallium arsenide is very dependent on crystal growth costs. These could be much reduced by the CLEFT technique in which gallium arsenide films are repeatedly regrown on the same substrate. Meanwhile, the potential of gallium arsenide multijunctions is very great. Efficiencies as high as 40% no longer seem like a pipe dream. Perhaps 45% will be reached, with concentrator module efficiencies in the 30–35% range. These are phenomenal values and are of significance to the future of very low-cost PV electricity. Indeed, at these very high efficiencies, several other issues (e.g., land area requirements) are also reduced.

Besides the possibility of adopting CLEFT for making concentrator cells, two other items are of importance for reducing related costs:

use of larger reactors and ways to increase the utilization of very expensive feedstock gases. These are being addressed through ongoing process-development R&D. Similarly, all issues related to cell cost depend on the module concentration ratio: if concentrations of 1000 suns *or more* can be reached, many cost issues will be resolved even for expensive cells.

Lenses are a related area of research. Low-cost lenses introduce losses. Achieving 90% or more transmission will require better designs and top-surface antireflection coatings. Other module hardware also contribute significant costs. Heat spreaders, fins on the back to dissipate heat, and module assembly add significant costs. Larger production volumes and the adoption of robotics should reduce these costs greatly.

Stability is an issue with concentrators because of the unusual aspects of the system: very high levels of sunlight are reached, with concomitantly high thermal fluxes. For instance, we found that the point-contact silicon cell was not stable under very high concentrations of UV light, an effect only characteristic of high-concentration sunlight. The new gallium arsenide designs will have to undergo tests to establish their stability under the burden of even greater concentrations than those for silicon.

Finally, precise two-axis trackers introduce a degree of unreliability not present in simple flat-plate PV. O&M could be much higher. Reliability could be reduced. Yet the efficiency potential of concentrators is so great that even these added cost burdens should become quite acceptable. For use in the Southwest US deserts or other arid climates of the world, concentrators could become a very powerful technology. If methods of transmitting electricity (superconductors) or PV/hydrogen (pipelines) are adopted, these same regions could predominate as sites for PV. Land impacts would be minimized. The future of PV and of very high-efficiency concentrators (probably those related to gallium arsenide multijunctions) are intimately connected.

New Technologies

It would be a mistake to assume that some or several new PV technologies will not emerge. For instance, several other compound semiconductors for thin films could be developed (e.g., cadmium selenide, zinc phosphide, copper gallium selenide, zinc telluride), especially if polycrystalline thin-film multijunctions are developed. New *thin-film* crystalline silicon structures and low-cost processes could be investigated. Electrochemical cells and cells made from organic materials could be studied. High-efficiency multijunctions made from *single-crystal* copper indium diselenide and copper gallium sulfide–selenide alloys could be tried, as could the CdTe alloys with zinc or mercury. Thin-film, single-crystal gallium arsenide could be grown on single-crystal silicon modules. A means could be developed of recrystallizing polycrystalline thin films grown on low-cost substrates such as glass. Concentrators using mirrors to focus 10,000 suns could be developed, along with the cells needed to withstand such intense sunlight. Thermal/PV systems could be developed to use the wasted heat. PV cells could be used to absorb light energy lost during conventional combustion processes (e.g., natural gas combustion). Many of these ideas have been tried—often once, with little success—but increased interest in PV would give each of them a fairer chance.

In a way, this laundry list of new possibilities makes evident a theme that holds throughout PV: many, potentially successful options exist. The capabilities of PV to meet very ambitious cost goals are based on a generic rather than a material-specific quality. PV is a rich technology in which many approaches may prove to be successful.

The PV community should always be open to new PV materials and approaches; the past ten years (with the advent of a-Si, CIS, CdTe, GaAs) should be the best teacher on that score. After all, as far as theory is concerned, PV of 60% efficiency is possible, and someday, someone will achieve it.

Conclusions

Research progress has brought PV to a new level. PV in the range of 10–20 cents/kWh is expected in the 1990s. Avenues for further improvement exist and will also be exploited during the same decade. However, significant technical issues remain for almost all PV technologies. Much ongoing industrial and governmental R&D commitment—funding—will be needed to allow the technologies to reach their potentials. Yet, new technical possibilities wait in the wings, demonstrating that we have not gone much beyond the surface as far as technical options are concerned. Because of the documented progress of PV R&D, and the generic capability of the different PV approaches to make further progress, the likelihood of PV achieving its technical goals is very high.

17 ☼ The Future
Policy Issues

Three groups have crucial roles to play in the future of PV: the industry that manufactures PV, the governments that want to implement it, and the utilities that will use it or compete against it. Part of the concern raised in this book is the future role of US industry and the US government: will PV be a domestic economic force, or will we be importing it, the way we do so many other energy and high-technology items?

Industry

The history of the PV industry has been fraught with difficulties. The basic problem has been simple:

☐ PV systems cost too much to be competitive for most familiar uses.

☐ The markets in which PV has been able to compete were either foreign or distant from traditional distributors, or both.

☐ PV manufacturing capacity exceeded demand until 1989.

Thus, several manufacturers have been competing for small, difficult markets. From a purely business standpoint, this is a prescription for catastrophe. Yet there have been reasons for perpetuating it. Although the PV industry as it stands is miniscule in relation to the

vill make and sell PV in the 21st century, it does form a
hicle for the ongoing development of new PV technologies.
.... is its most important role: there is no more effective agent for
technical progress than a viable industry. In addition to this critical
role, the industry also allows for the introduction of PV for various
uses, demonstrates its reliability, and helps educate the public as to its
potential value. In terms of the public good, a viable US PV industry is
essential for the nation if we are to have a major role in PV manufacturing in the 21st century. US jobs and money are on the line.

Right now, the PV industry is especially vulnerable to corporate
problems. Government policies during the 1980s almost completely
deemphasized PV, giving investors a very negative message as to
government's potential support. Generally, concerns over energy supplies or environmental stresses were minimal all through the 1980s.
Understandably, the PV industry's losses throughout that period ham-
pered the raising of new capital. Now, when new manufacturing facilities would reduce costs significantly, new investment is hard to come
by. Companies providing new technical options have almost been shut
out of the financial community. Even companies as exciting as Photon
Energy—with their uniquely low-cost approach to cadmium telluride—have found it very hard to raise the small amount of capital
they need to begin production. In the meantime, the oil companies that
invested in PV during the oil crisis—as a hedge and perhaps to generate some good publicity—have been getting out. Table 17-1 is a snapshot of a small segment of the troubled PV industry.

This snapshot is of just one portion of the PV industry—but an
important one. It shows how desperate the issue has become. Most or
all of the companies in Table 17-1 could be owned by foreign interests
within the next two years.

ARCO Solar

A perfect example of the chaos in the US PV industry is the sale
of ARCO Solar to a German company, Siemens. ARCO Solar has been
the biggest US producer of PV and is perhaps the world leader in new,
thin-film technologies (CIS and a-Si). In particular, it almost totally
dominates CIS, holding all the world records for efficiency at all sizes

Table 17-1. The Troubled Polycrystalline Thin-Film Industry

Group	Material	Status
ARCO Solar	CIS	Sold to Siemens
Boeing	CIS	Seeking to sell technology
ISET	CIS	Very small (6 employees) and no capital for expansion
Photon Energy	CdTe	Very small (11 employees) and no capital for expansion
Ametek	CdTe	Selling technology (foreign buyers possible)
Chronar	CIS	Starting CIS; troubled financially
Solarex	CIS	Starting CIS; oil company (AMOCO) subsidiary

from cells through modules. Yet Atlantic Richfield decided to sell its solar subsidiary. No doubt, the company was disappointed with the time it took to develop the new technologies. But it must have made its decision in about 1986 on the basis that oil prices stabilized in the 1980s. PV was a hedge against rising oil prices for Atlantic Richfield, and when it perceived less need for that hedge, it dropped PV.

Is anyone certain that oil prices will not rise significantly again? After all, the US position of depending on imports is worse than ever. In the meantime, environmental issues, from the Alaskan oil spill to the greenhouse effect, are crowding the stage with concern over fossil fuels. Yet Atlantic Richfield did not keep the best PV company in the world. Sales of ARCO Solar during its final year increased over 30% and the company showed a small profit, its first ever. Progress in CIS occurred at ARCO Solar after 1986, about the time the decision to sell was made. Does anyone "downtown" know about CIS — arguably the most important PV technology in the world? Can Atlantic Richfield be accused of acting like the dinosaurs from whose remains it refines its feedstock oil? Is its miniscule brain having difficulty communicating with its muscle-bound limbs? Who are its advisors on *energy and environmental policy*? Everywhere else, consciousness of the value of PV is growing. But not among those at Atlantic Richfield who gave up on ARCO Solar.

But the story does not end there. US industry could have bought ARCO Solar. Instead, a West German company, Siemens, did. Siemens had a reason to be interested, of course: Chernobyl raised energy/environment issues in Europe several years ago, but the US was too far from the blast to take much notice. Then the greenhouse effect added to the Germans' interest. The German government has now allocated more money to PV R&D than the US government, and they see PV and PV/hydrogen as central to their energy policy.

But the US industry and investment climate are not yet energized by such issues. Previous losses in PV, and short-term planning horizons, keep US corporations from investing in an unproven technology like PV.

Why should US corporations invest in such risky work as developing a new, high-technology option? Instead, they can get government military contracts and earn margins that are three times greater and *guaranteed*. If the political groups who identify with free-market competition were really interested in the culprit that holds back US competitiveness, they would soon find it in the bulging military budget which (1) inflates the deficit and therefore raises cost of capital for businesses and (2) pays contractually guaranteed profits above anything that can be reached through an honest day's work—i.e., through real, free-market competition. So why should we be surprised that a West German company, burdened by neither high interest rates nor the siren song of easy pickings with the military, outbids us for new technology?

The best thing that can be said is that the era of huge military budgets is ending. Companies that have not been developing new consumer products because they could make money more easily by contracting with the government will now have to consider risk/reward ratios that are nonzero. Maybe companies like Boeing, McDonald Douglas, General Electric, Westinghouse, TRW, Rockwell, and others will once again begin developing nonmilitary commercial products.

Government

One could say that most of the problems faced by US industry are caused by government policies. But, of course, government is just the expression of its people. So the crux of the problem really resides in people's attitudes. The short-term, selfish attitudes of the 1980s have been devastating, and the government has represented them well.

All during the 1980s, the US government virtually ignored energy and environmental issues. Solar budgets dropped from over $600 million annually to under $100 million. The Reagan administration took the lead, and Congress did not voice many objections. Every new year brought a budget request that was half of the previous year's. Only during the latter half of the decade did resistance to the trend stiffen. Congress responded by holding the line against further cuts. Meanwhile, hope for greater stimulus—tax credits or low-interest loans—was almost nil.

The position of PV within the key government agency—the Department of Energy—has traditionally been that of an outsider. Concerns at the Department are centered in other areas: nuclear weapons production and nuclear energy. If PV is considered at all, it is viewed as a competitor to these interests. It is also seen as a bastion of those who oppose nuclear energy, and thus becomes the target of those who would like to see nuclear prosper. During the 1980s, when energy issues were being underplayed, the various forces who were antipathetic to PV were able to dominate budget decisions and hammer down PV funding.

But things are changing at DOE. There is a new, more open administration. Having an energy policy is again seen as something of value. Environmental issues are getting more attention within the Administration. Fairness in evaluating nonconventional energy choices is being developed. Surprisingly, it may be government that changes before industry. With the help of various active legislators (Tim Wirth, Al Gore, Claudine Schneider, Vic Fazio, and others), as well as shifts at DOE, new PV policies may come into effect and change the atmosphere in the US. We may see new stimulus for the US to recapture world leadership in PV. With more support from government, the industry/investment climate would no doubt improve significantly.

New Policies

The US government can do a lot to stimulate greater progress in PV and to improve the investment climate for PV companies. The possible policy choices are of the following kinds:

1. R&D funding
2. Low-interest loans
3. Tax credits
4. EPA regulations

5. Purchase of large projects
6. Changes in utility-related regulations

1. R&D Funding

Perhaps the least controversial and certainly the least costly policy would be to return the R&D program to a level consistent with its importance. Very few options for spending PV money have higher leverage than R&D money, especially when such money is in very tight supply. During the last decade, few advances in PV have occurred at companies without at least some proportion of their R&D funding coming from DOE. Indeed, some PV technologies have survived and flourished only because of such external funds. Their internal management would have deleted them from the balance sheet otherwise. Several key options—CIS, CdTe, a-Si, GaAs—either would not exist or would be at preliminary levels if not for DOE support. Yet even today, some of these are still supported at levels that can only be called pitiful: the total annual expenditures for CIS are about $4 million. The level for CdTe is even smaller. Very high-efficiency GaAs cells are being supported at about half that level. Even work in crystalline silicon is minimal. In fact, every aspect of PV continues on a shoestring at the existing $35 million level being provided by DOE and Congress.

A balanced, effective program at the $100 million annual level would certainly accelerate PV toward its ambitious cost goals. It would also help to train the new scientists who will be part of the industry when it is as mature and profitable as the semiconductor and computer industries are today.

We may feel confident that positive adjustments of the PV R&D budget are in the offing. Although the 1990 DOE budget is the same as the 1989 budget (a carryover from the last administration), a national energy policy plan is being developed by DOE. It should reveal the value and potential of PV and the DOE research program. We should expect to see a change by the 1991 budget.

Figure 48 shows a projection of the likely costs of intermittent AC PV for two scenarios: a baseline R&D program equal to about $36 million, and an accelerated one of $100–$150 million. This projection was developed by SERI and Sandia representatives as part of work on

DOE's national energy strategy. These costs are conservative in at least one sense: if a particular PV technology or company breaks away from the field, the cost goals may be achieved a lot sooner than Figure 48 would imply. Figure 48, like any consensus, assumes neither a worst-case nor best-case scenario.

2. Low-Interest Loans

PV is very different from most conventional forms of energy production in that it is very capital-intensive but requires no costly fuel. Hydroelectric power has similarities. Such projects as hydroelectric appear very costly when they are built. But existing hydroelectric facilities are prized by all who have them because they were built and paid for long ago. Inflation has made existing hydro almost "too cheap to meter," to lift a line from the early history of nuclear power. Hydroelectric is the lowest cost US source of electricity.

But PV arrays will be much the same. Their costs will be almost all capital costs. Once paid for, they will eventually come to seem very cheap—almost free. And there is no real reason PV cannot last more than the projected 30 years. The owner of a large PV system will become very rich in its 31st year.

One critical factor—absent from most utility regulatory meetings—is that PV systems are not vulnerable to unexpected fuel escala-

Figure 48. A projection of the likely costs of intermittent AC PV for three scenarios: a baseline DOE R&D program equal to about $36 million annually; an accelerated one of $100–$150 million; and a market-stimulation approach based on giving PV a 2 cents/kWh premium versus other methods of making electricity that are environmentally more stressful. With greater funding and/or individual successes in private industry, PV costs could be reduced at an even greater rate than these projections.

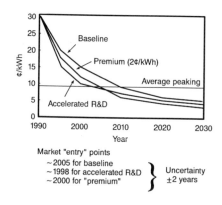

tions. If there is another oil supply crisis, the consumer will pay a lot more for fuel-based electricity. The consumer bears all the risk, because such costs are automatically passed through from the utility to the rate-payer. No such favor is given to a high-capital energy option.

The price of capital determines the price of PV. In fact, were PV to be built today with the same interest rates that prevailed when most hydro was built (in the 1930s at 3% interest or below), *it would be competitive with conventional electricity right now.* PV costs are roughly proportional to the interest rate. Current costs of 30 cents/ kWh are based on existing loan rates: 9.1% for utilities (the lowest of any in the corporate community). At 3% rates (one third of present rates), the exact same PV array would cost about 10 cents/kWh—a competitive cost for peaking power. Even more importantly, the next generation of PV—now on the drawing board—which will cost 10–15 cents/kWh at current rates would cost 3–5 cents/kWh at 3% interest. Those at 4 cents/kWh would be closer to a penny.

If the federal government were to raise capital for large PV projects with tax-free bonds, and if it were to subsidize interest rates for PV, it would have PV projects that would make cost-effective electricity today. The subsidy could be rationalized as payment for PV's reduced environmental stresses. Twenty years from now, when everything fuel-driven had ascended in cost with inflation, PV would be providing very inexpensive electricity.

However, any subsidy for PV—whether low-interest loans or tax credits—should be done with special criteria guaranteeing the money is being used to support technological acceleration. For example, to receive funding, costs of the total PV system should pass some test. Each year, the government should pay less for a new PV system. Prices would drop at some reasonable rate for every new project. If systems could not meet these cost criteria, support for new projects would be suspended until they do. This would keep manufacturers on their toes—a much to be desired state.

3. Tax Credits

Tax credits are another form of subsidy for PV projects. They work best, however, when the business is profitable, which is not the case with PV. Tax credits could be used with low-interest loans to

If we do not pursue one or more of these options, there is a good chance that foreign interests will threaten the US position in PV. Our companies are very vulnerable to foreign capital because US capital remains very expensive and hard to obtain. The government can start changing that condition; or it can stand by and watch its existing investment in PV—over a billion dollars—form the foundation of foreign domination.

18 ☀ The Challenge

PV could be a major provider of the world's electricity during the 21st century. It has that potential. But what we do today will determine the timing of PV's large-scale use. It will also determine whether the US will play an active role in PV. In a way, the questions surrounding PV are the same as those we must ask ourselves about so many other aspects of our nation's future:

☐ Can we sustain an attitude of investing in our future rather than merely responding to the crises of the present?

☐ Are we to be a "Third World" country, content to use up our vast natural resources? Or will we have the commitment to science and education needed to create high-technology, high-value products?

The problems we have with developing PV are very similar to those we face with many other high technologies. We are losing across-the-board in most new technologies. We still have the technical talent to develop most of them, but we give up on them (or are outbid) when it comes time for a significant investment in taking them to the marketplace. It has been a cliche that the US invents new technologies, then Japan or Germany markets them. That is as true of PV as it is of anything else: Bell Labs invented the silicon cell and made the first CIS cell but does no work in PV; RCA holds the patents for a-Si but does no work in PV (and Japan and Germany are on a par with us in a-Si); Boeing, DOE, SERI, and Atlantic Richfield have dominated copper indium diselenide, but soon Siemens owns most of it.

But even our ability to dominate the *research* of new products is slipping away: we are losing the research teams themselves. Examples: Ametek Inc., which makes a substantial profit, cut its R&D staff by half in 1989 and put its CdTe group up for sale; Boeing developed CIS and led the world for 5 years without doing anything to commercialize—now they are selling their technology to the highest bidder. SOHIO purchased a small US company, Monosolar, to develop CdTe PV; but after British Petroleum's purchase of SOHIO, research was transferred to England, and no SOHIO research in CdTe continues.

The general situation of new product R&D has gotten worse, not better, as corporations maximize their next quarter's profits, cutting R&D as "fat" along the way. Even our historic lead in the research of new products is fast disappearing. If the trend continues, we will be unable to develop competitive high-tech products. We will be frozen out. Our computers will be like Soviet computers. Our young people will have few choice job opportunities. We have already given away the lead—and any near-term chance of getting back in the race—in televisions, stereos, VCRs, and automobiles.

Clearly, the problem of developing PV is larger than any policy that is just concerned with PV. We can (and should) accelerate PV R&D; we can and should develop market stimulation policies (low-interest loans, tax credits, environmental regulations). But no matter what we do, we will be working within an ongoing context of a country without long-term ambitions or long-term planning; without a commitment to education; without an investment climate capable of sustaining long-term risks and rewards. On top of this, we have the huge Reagan budget deficit to work off for the next 40 years; and the remainder of the resource drain of the enormous military budget—not to mention other ills such as our accelerating medical inflation.

No matter what we do, no matter how successful our research program in PV, we may find that our technologies are bought out from under us by others who are capable of greater risk-taking—or merely have a lower cost of money (lower interest rates). At least half of the US PV infrastructure—small as it is—has already succumbed to this process of being purchased by foreigners. The remaining US companies cannot be expected to turn down foreign capital. If they did, they might not survive while they awaited US investors.

This does not mean that every PV company is a gem, or that every one of them will eventually make its investors rich. Investors must be able to live with uncertainty in order to gain great future payoffs. It means that we must have the patience, strength of character, and confidence that a few companies will make it. By doing so, those companies will pay us handsomely in jobs, profits, and hope for the future.

Scenarios for Improvement

We cannot solve the world's problems. Nor can we expect to markedly change the attitudes of our own country. But we can try a few things that might help the development of US PV. Maybe those actions would teach us something that would aid us to have better attitudes about the development of other high technologies in this country.

For PV, the highest priority of all is to increase the PV R&D budget to a reasonable level.

Priority One:
A $100 Million Annual PV R&D Budget

Right now, the DOE PV budget is about $35 million, which includes all the government-sponsored R&D in crystalline silicon, gallium arsenide, concentrators, copper indium diselenide, cadmium telluride, amorphous silicon, and new materials for industry and universities. It includes also all PV systems design and analysis work, module reliability testing, health, safety, and environmental studies, and all educational efforts. The companies and universities in PV are barely surviving, new scientists are not being trained to become part of the PV technical community, progress in scaling-up critical new technologies has been slowed by extremely tight resource limits, and almost all work in very new technologies (new materials, new structures) has ceased. The existing options are barely surviving, while new options are not even being considered.

A *low* level for a reasonable federal budget commitment would be $100 million annually for about the next 10 years. And this money should be clearly earmarked for research, not demonstration projects.

If and only if the basic $100 million annual R&D budget is provided, then demonstrations and the like can be considered.

The model for wisely using the extra R&D budget should be government–industry partnerships, i.e., direct support of the US PV companies that are committed to developing new technologies and commercializing them. This approach, at very low levels ($0.5–$1.5 million/year), has proved itself successful in stimulating rapid technical progress. It should be continued and expanded. Much of the R&D work will be process research, to find new, lower-cost film deposition methods, to improve yields, and to design production lines capable of immense production. University research teams should be funded as support groups to the industrial representatives. The national laboratories (e.g., SERI, Sandia, Institute of Energy Conversion) should also have a secondary but important support role supporting industry with fundamental research.

Besides filling out the funding levels of the existing technical programs, some new money should be allocated to: the investigation of PV with storage or hydrogen production and the investigation of new PV materials. A program devoted to supporting nonmainstream, high-risk areas—concentrators using mirrors rather than lenses, recrystallization of low-cost thin films, etc.—should also be opened and given enough money to get off the ground.

However, government must recognize that its role cannot stop with research funds. Government is trying to stimulate a new technology. If it uses industry as a vehicle for that technology, it must have a market for them. Companies that are chosen by government for support because they have expressed a commitment to commercialize PV must have markets. In a way, the existing, successful government program is at least partially responsible for the continued problems of the PV industry: the federal program stimulates new competitors and the new technologies they sell. How can the government create an industrial commitment for some future period without assisting the companies through the time in which their product is not fully competitive? Even if PV progress toward low cost continues unabated, the industry may remain only marginally profitable (or worse) for the next 10 years.

The same problem is associated with the loss of US firms to foreign investors. US investors cannot yet tolerate the risks associated with PV. Government must offset that risk without deadening the firms it subsidizes.

Priority Two:
Environmental Regulations/Low-Interest Loans

Perhaps the key to the impact of PV in the next 20 years (i.e., before it is clearly a least-cost alternative) will be environmental issues—and environmental regulation. Today the issue is the Clean Air Act; tomorrow it may be legislation to control the greenhouse gas problem. In every case, PV could be mandated to replace or supplement existing fossil and nuclear facilities as a response to these environmental concerns.

One of the least painful approaches to achieving this substitution would be to require that utilities meet certain environmental standards. If they exceed these clean air/CO_2 standards, they must either install scrubbers, pay a tax, or add nonpolluting generation like PV. This is the same concept as is now being proposed for the Clean Air Act in terms of sulfur dioxide emissions from utilities (although no PV is being included as yet). One way or another, the funds generated by such legislation could support the transition to cleaner alternatives like PV.

By mandating a regulatory climate in which new technologies are incorporated, the government would establish *a PV market worthy of its R&D program*. Of course, standards should be established mandating that the money be spent on US-made PV products, not on imports.

Such a program is not without pitfalls. We know that a great deal of money was wasted on demonstration projects in the late 1970s and early 1980s. The problem was twofold: (1) the R&D budget was not adequate or directed at emerging technologies, and (2) the large amounts directed to market stimulation were not carefully designed to accelerate progress in PV. PV systems that were not the least costly were purchased; last year's technology was perpetuated. In fact, decision-makers of the time should have realized that PV was not ready for demonstration. The preponderance of money should have been spent on R&D—and R&D for innovative options, at that.

If regulations to mandate the purchase of PV are set up, they should include an ambitious requirement for cost reductions. Each year, the price paid for PV should be reduced, with the aim of reaching a cost-competitive level by about 2010. In the past, market stimulation programs have been severely criticized for stimulating the sale of last-year's technology. Only by avoiding this stigma can such a project succeed and gain the confidence of the public. *We do not want PV to get the reputation of being a noncompetitive luxury forced down our throats by the government.* Stiff cost reductions should help us avoid this catastrophe. On the other hand, the public will have to be educated to the process and allowed to see its long-term payoff.

In the 1990s, the government should provide low-interest loans and mandate the purchase of PV. The purchases themselves should be environmentally driven. In that way, it should be clear that PV systems have added value over conventional energy sources. Awareness of PV's environmental value will at least partially offset the negative image of the higher cost of PV.

Fortunately, we do not need many billions of dollars worth of PV projects to stabilize the US PV industry. In fact, it would probably be a mistake to raise the amounts spent for these projects too rapidly. If we spend even small amounts, industry will be encouraged. Later, when PV becomes a better buy, potential buyers such as industry, munici-palities, federal agencies, and others should be forced to include PV within their mix of electrical generation.

The requirement for annual cost reductions should be the center-piece of this effort. Without it, the American people would have every right to claim that an inefficient, expensive technology was being foisted on them. If in any particular year, the planned cost reductions were not met, that should not be regarded as a catastrophe. If some companies go bankrupt because they do not stay on the leading edge, that should also be no problem. As long as progress is maintained over

Table 18-1. Projected PV Electric Utility Market in the US

	Cumulative (GW_{peak})[a]					
	1990	1995	2000	2010	2020	2030
Baseline	0.01	0.05	1	6	48	160
Accelerated R&D	0.01	0.15	5	80	240	480[b]
"Premium"	0.01	0.10	2	16	160	480[b]

[a]2 $GW_{peak}PV \sim 1GW$ (supply, conventional). 40 $GW_{peak}PV \sim 1$ quad (primary equivalent).
[b]Approximately 25% of cumulative capacity need in 2030; assumes storage.

the long haul, and the public sees that they are nurturing a valuable new energy option, the program will be a success.

Table 18-1 shows a view of the potential of PV to make a reasonable contribution to our energy mix by 2030. The baseline case is business as usual. Even with business as usual, we may assume that someone—the Japanese, the Germans, possibly someone in the US—will eventually succeed in making inexpensive PV. The 160 GW_p in 2030 (baseline case) would just be the beginning. The other cases are "accelerated research, development, and demonstration" (a $100–$150 million annual DOE PV budget) and an "environmental premium" (2 cents/kWh). In those cases, progress is faster and the size of the penetration is bigger by 2030. Even these more optimistic markets are conservative projections in the sense that they only assume PV is to be used for electricity production—no PV to provide hydrogen fuel or electricity for battery-powered vehicles. Storage is assumed for contributions that are larger than 10% of the grid capacity, which would be the case in the 2025 period. By then, storage of reasonable cost should be available. A capacity of about 500 GW_p in 2030 would be equivalent to about 10^{12} kWh of electricity, a substantial contribution and one that would no doubt be healthy from the standpoint of reducing environmental stresses. It would result in the displacement of about 250 billion tons of carbon dioxide annually.

Final Word

People are anxious to know that there are practical alternatives to conventional fossil and nuclear fuels. Even today, major meetings concerned with ameliorating the greenhouse effect can be held without mention of PV or the other solar technologies. But the future of PV does not really depend on the US position: someone will do it, even if we do not. Only the question of timing and who will dominate the technology are at stake.

The real issues for PV are technical ones, and their solutions depend on resource commitments. PV must continue to overcome technical barriers. As it does so, its cost will continue to fall. Based on existing technical approaches covered in this book, we can easily foresee PV costing under 10 cents/kWh. We can claim that there are clear paths to even lower costs—say 3–5 cents/kWh. But the horizon for PV has not yet been reached. We are just beginning to develop this technology. It has the promise of efficiencies far in excess of those that we are used to. Similarly, the true cost limits could be well below those we assume, since the active, semiconductor materials going into these devices can cost as little as pennies per square meter. These issues will be more fully addressed in the next century. Meanwhile, we as a nation have the challenge to develop PV. By rising to that challenge, we can expect to reap its many benefits. We can have a major, new, environmentally sound method of producing a large portion of our energy. And we can do it ourselves, rather than buy it from someone else.

☼ Index